W9-BZW-831

PERIODIC TABLE OF THE ELEMENTS

Lanthanides and Actinides

PERIODIC TABLE OF THE ELEMENTS

Lanthanides and Actinides

Monica Halka, Ph.D., and
Brian Nordstrom, Ed.D.

LANTHANIDES AND ACTINIDES

Facts On File, Inc.
An imprint of Infobase Publishing
132 West 31st Street
New York NY 10001

Library of Congress Cataloging-in-Publication Data
Halka, Monica.
Lanthanides and actinides / Monica Halka and Brian Nordstrom.
p. cm.—(Periodic table of the elements)
Includes index.
ISBN 978-0-8160-7372-6
1. Rare earth metals. 2. Actinide elements. 3. Periodic law. I. Nordstrom, Brian. II. Title.
QD172.R2H245 2011
546'.41—dc22 2010006296

Facts On File books are available at special discounts when purchased in bulk quantities for businesses, associations, institutions, or sales promotions. Please call our Special Sales Department in New York at (212) 967-8800 or (800) 322-8755.

You can find Facts On File on the World Wide Web at http://www.factsonfile.com

Text design by Erik Lindstrom
Composition by Hermitage Publishing Services
Illustrations by Richard Garratt and Sholto Ainslie
Photo research by Tobi Zausner, Ph.D.
Cover printed by Bang Printing, Brainerd, Minn.
Book printed and bound by Bang Printing, Brainerd, Minn.
Date printed: December 2010
Printed in the United States of America

10 9 8 7 6 5 4 3 2 1

This book is printed on acid-free paper.

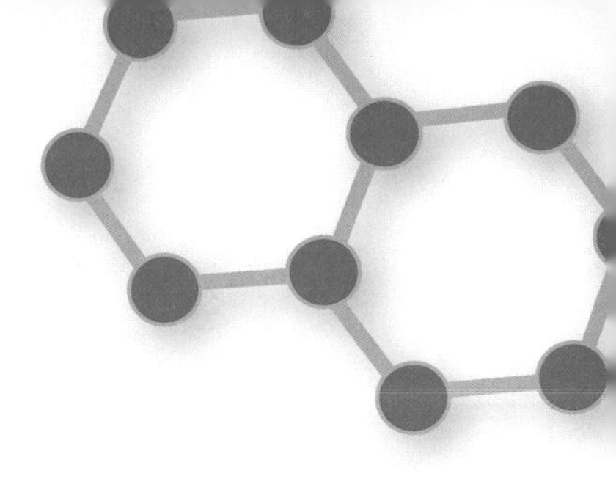

Contents

Preface

Speculations about the nature of matter date back to ancient Greek philosophers like Thales, who lived in the sixth century B.C.E., and Democritus, who lived in the fifth century B.C.E., and to whom we credit the first theory of *atoms.* It has taken two and a half millennia for natural philosophers and, more recently, for chemists and physicists to arrive at a modern understanding of the nature of *elements* and *compounds.* By the 19th century, chemists such as John Dalton of England had learned to define elements as pure substances that contain only one kind of atom. It took scientists like the British physicists Joseph John Thomson and Ernest Rutherford in the early years of the 20th century, however, to demonstrate what atoms are—entities composed of even smaller and more elementary particles called *protons, neutrons,* and *electrons.* These particles give atoms their properties and, in turn, give elements their physical and chemical properties.

After Dalton, there were several attempts throughout Western Europe to organize the known elements into a conceptual framework that would account for the similar properties that related groups of elements exhibit and for trends in properties that correlate with increases in atomic weights. The most successful *periodic table* of the elements was designed in 1869 by a Russian chemist, Dmitri Mendeleev. Mendeleev's method of organizing the elements into columns grouping elements with similar chemical and physical properties proved to be so practical that his table is still essentially the only one in use today.

While there are many excellent works written about the periodic table (which are listed in the section on further resources), recent scientific investigation has uncovered much that was previously unknown about nearly every element. The Periodic Table of the Elements, a six-volume set, is intended not only to explain how the elements were discovered and what their most prominent chemical and physical properties are, but also to inform the reader of new discoveries and uses in fields ranging from astrophysics to material science. Students, teachers, and the general public seldom have the opportunity to keep abreast of these new developments, as journal articles for the nonspecialist are hard to find. This work attempts to communicate new scientific findings simply and clearly, in language accessible to readers with little or no formal background in chemistry or physics. It should, however, also appeal to scientists who wish to update their understanding of the natural elements.

Each volume highlights a group of related elements as they appear in the periodic table. For each element, the set provides information regarding:

- the discovery and naming of the element, including its role in history, and some (though not all) of the important scientists involved;
- the basics of the element, including such properties as its atomic number, atomic mass, electronic configuration, melting and boiling temperatures, abundances (when known), and important isotopes;
- the chemistry of the element;
- new developments and dilemmas regarding current understanding; and
- past, present, and possible future uses of the element in science and technology.

Some topics, while important to many elements, do not apply to all. Though nearly all elements are known to have originated in stars or stellar explosions, little information is available for some. Some others that have been synthesized by scientists on Earth have not been observed in stellar spectra. If significant astrophysical nucleosynthesis research exists, it is presented as a separate section. The similar situation applies for geophysical research.

Special topic sections describe applications for two or more closely associated elements. Sidebars mainly refer to new developments of special interest. Further resources for the reader appear at the end of the book, with specific listings pertaining to each chapter, as well as a listing of some more general resources.

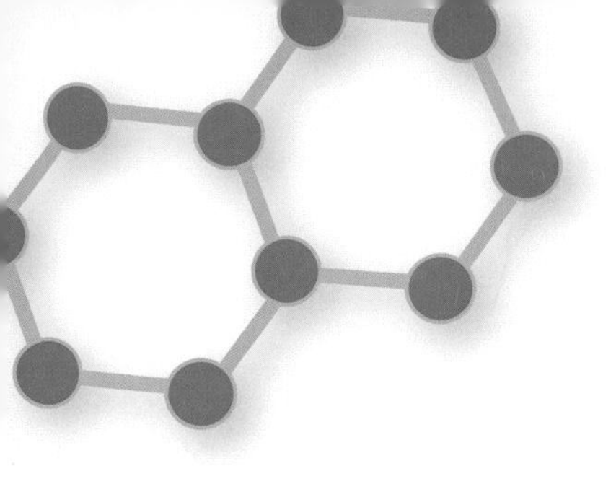

Acknowledgments

First and foremost, I thank my parents, who convinced me I was capable of achieving any goal. In graduate school, my thesis adviser, Dr. Howard Bryant, influenced my way of thinking about science more than anyone else. Howard taught me that learning requires having the humility to doubt your understanding and that it is important for a physicist to be able to explain her work to anyone. I have always admired the ability to communicate scientific ideas to nonscientists and wish to express my appreciation for conversations with National Public Radio science correspondent Joe Palca, whose clarity of style I attempt to emulate in this work. I also thank my coworkers at Georgia Tech, Dr. Greg Nobles and Ms. Nicole Leonard, for their patience and humor as I struggled with deadlines.

—Monika Halka

In 1967, I entered the University of California at Berkeley. Several professors, including John Phillips, George Trilling, Robert Brown, Samuel Markowitz, and A. Starker Leopold, made significant and lasting impressions. I owe an especial debt of gratitude to Harold Johnston, who was my graduate research adviser in the field of atmospheric chemistry. Many of the scientists mentioned in the Periodic Table of the Elements set I have known personally: For example, I studied under Neil Bartlett, Kenneth Street, Jr., and physics Nobel laureate Emilio Segrè. I especially cherish having known chemistry Nobel laureate Glenn Seaborg. I also acknowledge my past and present colleagues at California State University, Northern Arizona University, and Embry-Riddle Aeronautical University, Prescott, Arizona, without whom my career in education would not have been as enjoyable.

—*Brian Nordstrom*

Both authors thank Jodie Rhodes and Frank Darmstadt for their encouragement, patience, and understanding.

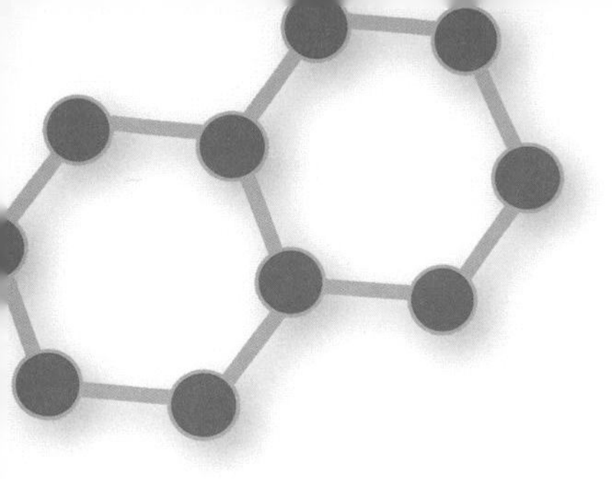

Introduction

Most discussions of the chemical elements include substances that are familiar to the general public. Elements such as hydrogen, oxygen, helium, chlorine, carbon, sodium, calcium, aluminum, and iron, and many others, are discussed in everyday conversation. In this book, with the exception of uranium and plutonium (two of the actinides), that familiarity is probably not the case. Few people have ever heard of most of the lanthanides. Consider, for example, praseodymium, terbium, dysprosium, holmium, or lutetium. The lanthanides are unknown to most people because these elements have few familiar uses (unlike aluminum, iron, or silicon), nor are they important in the biochemistry of plants or animals (unlike carbon, oxygen, sodium, or calcium, which are essential elements). The lanthanides and actinides do have important uses, which are described below. With the exception of people who work in industries that employ these elements, most people are unaware of them. Even students who have taken chemistry in high school or college are unlikely to know anything about the lanthanides except for their position in the periodic table. It would be very unusual for students to perform experiments with any of the lanthanides; for one thing, the cost of lanthanide compounds would be beyond most schools' budgets.

In the case of the actinides (and transactinides), unfamiliarity is largely due to the fact that these elements are all radioactive. Most of them do not even occur naturally on Earth but have to be produced artificially. In several cases, only a few atoms of the heavier elements have ever been made, and almost nothing is known about them. Uranium is an exception because of its prominence in both the nuclear power industry and in atom bombs. Although plutonium does not occur naturally, tons of it have been synthesized since its discovery, sometimes for use in nuclear weapons. (For reasons of national security, plutonium is not used in civilian nuclear reactors.) The names of several of the elements in this book, however, may in fact be recognizable to readers who have not studied science formally since many of the *transuranium elements* (elements heavier than uranium) have been named for famous people. Examples include curium (named for Pierre and Marie Curie), einsteinium (named for Albert Einstein), and nobelium (named for Alfred Nobel, the creator of the Nobel Prizes).

Despite the unfamiliarity the general public has with the lanthanides, actinides, and transactinides, these elements make up approximately 35 percent of the total number of known elements. Attempts to produce new elements—or new isotopes of known elements—constitute an active area of scientific research. In fact, the search for new elements has put the locations of three laboratories on the world scientific map: Berkeley, California; Darmstadt, Germany; and Dubna, Russia (together with the relatively recent addition of laboratories in Japan). Several Nobel Prizes have been awarded in both chemistry and physics to scientists who either discovered new elements directly or whose work contributed to their discoveries. Most notable in this regard is the late Berkeley nuclear chemist Glenn T. Seaborg (1912–99) for whom element 106 (seaborgium) is named. Seaborg's coworker, nuclear scientist Albert Ghiorso (1915–), is still an active researcher and is even listed in the *Guinness Book of World Records* for having participated in the discoveries of the most elements—all 12 elements from americium (number 95) through seaborgium. German chemists Sigurd Hofmann (1944–) and Peter Armbruster (1931–) at Darmstadt continue to extend the periodic table.

Lanthanides and Actinides will help fill the gaps in the public's knowledge. Readers will come to appreciate how rich the histories of the elements' discoveries are, as well as learn what the practical applications are that at least some of the elements have in today's scientific, technological, medical, and military communities.

The similarities among the lanthanides and the actinides are due to their electronic configurations. The lanthanides are the first elements in the periodic table in which "f" orbitals are populated with electrons. The occupancy of "f" orbitals with electrons is what sets lanthanides and actinides apart from the other elements and gives the name the *"f"-block* to the subunit of the periodic table occupied by these elements. In the case of the lanthanides, the "4f" subshell is being populated with electrons. In the case of the actinides, the "5f" subshell is being populated.

In the main body of the periodic table, the Group IIIA elements, in descending order, are scandium (Sc), yttrium (Y), lanthanum (La), and actinium (Ac). The properties of the 14 elements (the *lanthanides*) that follow lanthanum in order of atomic number are so similar to the properties of lanthanum that, effectively, they occupy the same place as lanthanum in the table. Likewise, the properties of the 14 elements (the *actinides*) that follow actinium effectively occupy the same place as actinium. Because it would be impractical to actually show all of those elements in the same places, the lanthanides and actinides are split out of the main body of the periodic table into their own rows.

One member of the lanthanides—element 61, promethium—does not occur naturally on Earth. All of promethium's isotopes are radioactive with extremely short half-lives. All the other lanthanides do occur naturally and tend to be found together in similar ores. Of the actinides, only thorium, protactinium, and uranium occur naturally (though evidence indicates that plutonium was produced in a natural underground reactor before recorded history; see chapter 3). Even so, they have no stable isotopes. There are isotopes of thorium and uranium with half-lives exceptionally long enough that primordial thorium and uranium still exist from the time of Earth's formation. Primordial protactinium no longer exists; however, small amounts are continuously being produced by the radioactive decay of uranium. All of the elements past uranium (the transuranium elements) must be produced artificially. After californium (element 98), only very tiny amounts of these heavier elements have ever been made. Quantities are too small for anyone to have ever actually seen these elements or their compounds with the naked eye.

The *transactinides* are the elements heavier than the heaviest actinide—lawrencium (element 103). These elements are not actually part of the "f" block of the periodic table. Elements 104–112 are transition metals, and elements 113–118 are post-transition metals or nonmetals. However, the elements that follow lawrencium have been included in this book because the production of transactinide elements follows the same methods that were used to produce the transuranium elements through lawrencium.

At the time this book is being written, no conclusive report has been made of element 117. In February 2010, the International Union for Pure and Applied Chemistry adopted the name copernicium (symbol Cn) for element 112 after the great Polish astronomer Nicolaus Copernicus (1473–1543). Names and symbols for elements 113–118 have not yet been proposed. Production of new elements, as well as production of new, perhaps longer-lived, isotopes of known elements, is, however, an ongoing research activity in California, Germany, Russia, and Japan. The International Union for Pure and Applied Chemistry (IUPAC) rules on claims of discovery and officially recognizes proposed names and symbols. As names and symbols are adopted, the periodic table will be updated to reflect the changes.

Chapter 1 covers the lanthanide or rare earth elements, which are becoming increasingly important in alternative energy technology. Rare earths are used in lasers, mercury street lights, and in glass coloration. The oxides of the rare earths have excellent abrasive properties and are used to polish mirrors and the glass surfaces of televisions, and chlorides of the rare earths are used as catalysts in the petroleum refining industry.

Chapter 2 explores actinium, thorium, and proctactinium. The main uses of actinium and protactinium are in scientific research, but thorium is a different story. Thorium-fueled fission reactors could, in the long term, provide a large fraction of the energy demand of the planet more efficiently and with fewer hazardous side effects than present-day uranium enrichment reactors. Currently, thorium oxide has several consumer applications, including its use in high-refractive-index glass and as a catalyst in petroleum and sulfuric *acid* production.

Chapter 3 is dedicated to the investigation of uranium, which has such a rich history and current importance. Uranium is mined for use in nuclear reactors and weapons. Nuclear fission reactors provide 14 percent of the world's electricity, 20 percent of electricity in the United States, and 100 percent of the plutonium used in nuclear weapons. The United States has more nuclear reactors in operation than any other country, but uranium production may not be able to keep up with demand. In research, collisions of uranium atoms can simulate hot, dense nuclear matter and assist scientists studying conditions shortly after the big bang.

Chapter 4 covers the 11 transuranium elements, of which plutonium and americium are the most useful. The most familiar use of americium is its application as the detecting chemical in smoke detectors. For research purposes, when combined with beryllium, americium also serves as a reliable source for neutrons. Plutonium is produced in significantly larger quantities than any of the other transuranium elements. It is used as fissile material in nuclear reactors and in the primary stage of modern thermonuclear weapons. Neptunium is a byproduct of plutonium production and has no major commercial uses. Curium is used in pacemakers and in power supplies for satellites. Any of the transcurium elements are produced in such minuscule quantities that they have no uses other than in scientific research, but they carry a rich and interesting history of discovery.

Chapter 5 explores the discovery and synthesis of the transactinides—the elements with atomic numbers greater than 103. While the transactinide elements are most likely synthesized in some very massive stars, astrophysicists have no way of observing this behavior because these elements are so short-lived, decaying to another element almost as soon as they are born. On Earth, the only way to synthesize the transactinides is in high-energy collisions. The discoveries of the transactinide elements have resulted from work done at three principal research laboratories—one in the United States, one in Germany, and one in Russia.

Chapter 6 presents possible future developments that involve the lanthanides and actinides.

Lanthanides and Actinides provides the reader, whether student or scientist, with an overall up-to-date understanding regarding the lanthanide, actinide, and transactinide elements—where they came from, how they fit into the current technological society, and where they may lead.

Overview: Chemistry and Physics Background

What *is* an element? To the ancient Greeks, everything on Earth was made from only four elements—earth, air, fire, and water. Celestial bodies—the Sun, moon, planets, and stars—were made of a fifth element: ether. Only gradually did the concept of an element become more specific.

An important observation about nature was that substances can change into other substances. For example, wood burns, producing heat, light, and smoke and leaving ash. Pure metals like gold, copper, silver, iron, and lead can be smelted from their ores. Grape juice can be fermented to make wine and barley fermented to make beer. Food can be cooked; food can also putrefy. The baking of clay converts it into bricks and pottery. These changes are all examples of chemical reactions. Alchemists' careful observations of many chemical reactions greatly helped them to clarify the differences between the most elementary substances ("elements") and combinations of elementary substances ("compounds" or "mixtures").

Elements came to be recognized as simple substances that cannot be decomposed into other even simpler substances by chemical reactions. Some of the elements that had been identified by the Middle Ages are easily recognized in the periodic table because they still have chemical symbols that come from their Latin names. These elements are listed in the following table.

ELEMENTS KNOWN TO ANCIENT PEOPLE

Iron: Fe ("ferrum")	Copper: Cu ("cuprum")
Silver: Ag ("argentum")	Gold: Au ("aurum")
Lead: Pb ("plumbum")	Tin: Sn ("stannum")
Antimony: Sb ("stibium")	Mercury: Hg ("hydrargyrum")
*Sodium: Na ("natrium")	*Potassium: K ("kalium")
Sulfur: S ("sulfur")	

*Sodium and potassium were not isolated as pure elements until the early 1800s, but some of their salts were known to ancient people.

Modern atomic theory began with the work of the English chemist John Dalton in the first decade of the 19th century. As the concept of the atomic composition of matter developed, chemists began to define elements as simple substances that contain only one kind of atom. Because scientists in the 19th century lacked any experimental apparatus capable of probing the structure of atoms, the 19th-century model of the atom was rather simple. Atoms were thought of as small spheres of uniform density; atoms of different elements differed only in their masses. Despite the simplicity of this model of the atom, it was a great step forward in our understanding of the nature of matter. Elements could be defined as simple substances containing only one kind of atom. Compounds are simple substances that contain more than one kind of atom. Because atoms have definite masses, and only whole numbers of atoms can combine to make molecules, the different elements that make up compounds are found in definite proportions by mass. (For example, a molecule of water contains one oxygen atom and two hydrogen atoms, or a mass ratio of oxygen-to-hydrogen of about 8:1.) Since atoms are neither created nor destroyed during ordinary chemical reactions ("ordinary" meaning in contrast to "nuclear" reactions), what happens in chemical reactions is that atoms are rearranged into combinations that differ from the original reactants, but in doing so, the total mass is conserved. Mixtures are combinations of elements that are not in definite proportions. (In salt water, for example, the salt could be 3 percent by mass, or 5 percent by mass, or many other possibilities; regardless of the percentage of salt, it would still be called "salt water.") Chemical reactions are not required to separate the components of mixtures; the components of mixtures can be separated by physical processes such as distillation, evaporation, or precipitation. Examples of elements, compounds, and mixtures are listed in the following table.

The definition of an element became more precise at the dawn of the 20th century with the discovery of the proton. We now know that an atom has a small center called the "nucleus." In the nucleus are one or more protons, positively charged particles, the number of which determine an atom's identity. The number of protons an atom has is referred to as its "atomic number." Hydrogen, the lightest element, has an atomic number of 1, which means each of its atoms contains a single proton. The next element, helium, has an atomic number of 2, which means each of its atoms contain two protons. Lithium has an atomic number of 3, so its atoms have three protons, and so forth, all the way through the periodic table. Atomic nuclei also contain neutrons, but atoms of the same element can have different numbers of neutrons; we call atoms of the same element with different number of neutrons "isotopes."

EXAMPLES OF ELEMENTS, COMPOUNDS, AND MIXTURES

ELEMENTS	COMPOUNDS	MIXTURES
Hydrogen	Water	Salt water
Oxygen	Carbon dioxide	Air
Carbon	Propane	Natural gas
Sodium	Table salt	Salt and pepper
Iron	Hemoglobin	Blood
Silicon	Silicon dioxide	Sand

There are roughly 92 naturally occurring elements—hydrogen through uranium. Of those 92, two elements, technetium (element 43) and promethium (element 61), may once have occurred naturally on Earth, but the atoms that originally occurred on Earth have decayed away, and those two elements are now produced artificially in nuclear reactors. In fact, technetium is produced in significant quantities because of its daily use by hospitals in nuclear medicine. Some of the other first 92 elements—polonium, astatine, and francium, for example—are so radioactive that they exist in only tiny amounts. All of the elements with atomic numbers greater than 92—the so-called transuranium elements—are all produced artificially in nuclear reactors or particle accelerators. As of the writing of this book, the discoveries of the elements through number 118 (with the exception of number 117) have all been reported. The discoveries of elements with atomic numbers greater than 111 have not yet been confirmed, so those elements have not yet been named.

When the Russian chemist Dmitri Mendeleev (1834–1907) developed his version of the periodic table in 1869, he arranged the elements known at that time in order of *atomic mass* or *atomic weight* so that they fell into columns called *groups* or *families* consisting of elements with

The Russian chemist Dmitri Mendeleev created the periodic table of the elements in the late 1800s. *(Science Museum, London/ The Image Works)*

similar chemical and physical properties. By doing so, the rows exhibit periodic trends in properties going from left to right across the table, hence the reference to rows as *periods* and name "periodic table."

Mendeleev's table was not the first periodic table, nor was Mendeleev the first person to notice *triads* or other groupings of elements with similar properties. What made Mendeleev's table successful and the one we use today are two innovative features. In the 1860s, the concept of *atomic number* had not yet been developed, only the concept of atomic mass. Elements were always listed in order of their atomic masses, beginning with the lightest element, hydrogen, and ending with the heaviest element known at that time, uranium. Gallium and germanium, however, had not yet been discovered. Therefore, if one were listing the known elements in order of atomic mass, arsenic would follow zinc, but that would place arsenic between aluminum and indium. That does not make sense because arsenic's properties are much more like those of phosphorus and antimony, not like those of aluminum and indium.

Mendeleev's Periodic Table (1871)

Period \ Group	I	II	III	IV	V	VI	VII	VIII
1	H=1							
2	Li=7	Be=9.4	B=11	C=12	N=14	O=16	F=19	
3	Na=23	Mg=24	Al=27.3	Si=28	P=31	S=32	Cl=35.5	
4	K=39	Ca=40	?=44	Ti=48	V=51	Cr=52	Mn=55	Fe=56, Co=59 Ni=59
5	Cu=63	Zn=65	?=68	?=72	As=75	Se=78	Br=80	
6	Rb=85	Sr=87	?Yt=88	Zr=90	Nb=94	Mo=96	?=100	Ru=104, Rh=104 Pd=106
7	Ag=108	Cd=112	In=113	Sn=118	Sb=122	Te=125	J=127	
8	Cs=133	Ba=137	?Di=138	?Ce=140				
9								
10			?Er=178	?La=180	Ta=182	W=184		Os=195, Ir=197 Pt=198
11	Au=199	Hg=200	Tl=204	Pb=207	Bi=208			
12				Th=231		U=240		

Dmitri Mendeleev's 1871 periodic table—the elements listed within are the ones that were known at that time, arranged in order of increasing relative atomic mass. Mendeleev predicted the existence of elements with masses of 44, 68, and 72: His predictions were later shown to have been correct.

To place arsenic in its "proper" position, Mendeleev's first innovation was to leave two blank spaces in the table after zinc. He called the first element *eka-aluminum* and the second element *eka-silicon,* which he said corresponded to elements that had not yet been discovered but whose properties would resemble the properties of aluminum and silicon, respectively. Not only did Mendeleev predict the elements' existence, he also estimated what their physical and chemical properties should be in analogy to the elements near them. Shortly afterward, these two elements were discovered and their properties were found to be very close to what Mendeleev had predicted. Eka-aluminum was called *gallium* and eka-silicon was called *germanium.* These discoveries validated the predictive power of Mendeleev's arrangement of the elements and demonstrated that Mendeleev's periodic table could be a predictive tool, not just a compendium of information that people already knew.

The second innovation Mendeleev made involved the relative placement of tellurium and iodine. If the elements are listed in strict order of their atomic masses, then iodine should be placed before tellurium, since iodine is lighter. That would place iodine in a group with sulfur and selenium and tellurium in a group with chlorine and bromine, an arrangement that does not work for either iodine or tellurium. Therefore, Mendeleev rather boldly reversed the order of tellurium and iodine so that tellurium falls below selenium and iodine falls below bromine. More than 40 years later, after Mendeleev's death, the concept of atomic number was introduced, and it was recognized that elements should be listed in order of atomic number, not atomic mass. Mendeleev's ordering was thus vindicated, since tellurium's atomic number is one less than iodine's atomic number. Before he died, Mendeleev was considered for the Nobel Prize, but did not receive sufficient votes to receive the award despite the importance of his insights.

THE PERIODIC TABLE TODAY

All of the elements in the first 12 groups of the periodic table are referred to as *metals.* The first two groups of elements on the left-hand side of the table are the *alkali metals* and the *alkaline earth metals.* All of the alkali metals are extremely similar to each other in their chemical and physical properties, as, in turn, are all of the alkaline earths to each other. The

10 groups of elements in the middle of the periodic table are *transition metals.* The similarities in these groups are not as strong as those in the first two groups, but still satisfy the general trend of similar chemical and physical properties. The transition metals in the last row are not found in nature but have been synthesized artificially. The metals that follow the transition metals are called post-transition metals.

The so-called *rare earth elements,* which are all metals, usually are displayed in a separate block of their own located below the rest of the periodic table. The elements in the first row of rare earths are called *lanthanides* because their properties are extremely similar to the properties of lanthanum. The elements in the second row of rare earths are called *actinides* because their properties are extremely similar to the properties of actinium. The actinides following uranium are called *transuranium elements* and are not found in nature but have been produced artificially.

The far right-hand six groups of the periodic table—the remaining *main group elements*—differ from the first 12 groups in that more than one kind of element is found in them; in this part of the table we find metals, all of the *metalloids* (or *semimetals*), and all of the *nonmetals.* Not counting the artificially synthesized elements in these groups (elements having atomic numbers of 113 and above and that have not yet been named), these six groups contain 7 metals, 8 metalloids, and 16 nonmetals. Except for the last group—the *noble gases*—each individual group has more than just one kind of element. In fact, sometimes nonmetals, metalloids, and metals are all found in the same column, as are the cases with group IVB (C, Si, Ge, Sn, and Pb) and also with group VB (N, P, As, Sb, and Bi). Although similarities in chemical and physical properties are present within a column, the differences are often more striking than the similarities. In some cases, elements in the same column do have very similar chemistry. Triads of such elements include three of the *halogens* in group VIIB—chlorine, bromine, and iodine; and three group VIB elements—sulfur, selenium, and tellurium.

ELEMENTS ARE MADE OF ATOMS

An atom is the fundamental unit of matter. In ordinary chemical reactions, atoms cannot be created or destroyed. Atoms contain smaller *subatomic* particles: protons, neutrons, and electrons. Protons and neutrons are located in the *nucleus,* or center, of the atom and are referred to as *nucleons.* Electrons are located outside the nucleus. Protons and neutrons are comparable in mass and significantly more massive than electrons. Protons carry positive electrical charge. Electrons carry negative charge. Neutrons are electrically neutral.

The identity of an element is determined by the number of protons found in the nucleus of an atom of the element. The number of protons is called an element's atomic number, and is designated by the letter *Z.* For hydrogen, Z = 1, and for helium, Z = 2. The heaviest naturally occurring element is uranium, with Z = 92. The value of Z is 118 for the heaviest element that has been synthesized artificially.

Atoms of the same element can have varying numbers of neutrons. The number of neutrons is designated by the letter *N.* Atoms of the same element that have different numbers of neutrons are called *isotopes* of that element. The term *isotope* means that the atoms occupy the same place in the periodic table. The sum of an atom's protons and neutrons is called the atom's *mass number.* Mass numbers are dimensionless whole numbers designated by the letter *A* and should not be confused with an atom's *mass,* which is a decimal number expressed in units such as grams. Most elements on Earth have more than one isotope. The average mass number of an element's isotopes is called the element's atomic mass or atomic weight.

The standard notation for designating an atom's atomic and mass numbers is to show the atomic number as a subscript and the mass number as a superscript to the left of the letter representing the element. For example, the two naturally occurring isotopes of hydrogen are written ${}^{1}_{1}H$ and ${}^{2}_{1}H$.

For atoms to be electrically neutral, the number of electrons must equal the number of protons. It is possible, however, for an atom to gain or lose electrons, forming *ions.* Metals tend to lose one or more electrons to form positively charged ions (called *cations*); nonmetals are more likely to gain one or more electrons to form negatively charged ions (called *anions*). Ionic charges are designated with superscripts. For example, a calcium ion is written as Ca^{2+}; a chloride ion is written as Cl^{-}.

THE PATTERN OF ELECTRONS IN AN ATOM

During the 19th century, when Mendeleev was developing his periodic table, the only property that was known to distinguish an atom of one element from an atom of another element was relative mass. Knowledge of atomic mass, however, did not suggest any relationship between an element's mass and its properties. It took several discoveries—among them that of the electron in 1897 by the British physicist John Joseph ("J. J.") Thomson, *quanta* in 1900 by the German physicist Max Planck, the wave nature of matter in 1923 by the French physicist Louis de Broglie, and the mathematical formulation of the quantum mechanical model of the atom in 1926 by the German physicists Werner Heisenberg and Erwin Schrödinger (all of whom collectively illustrate the international nature of science)—to elucidate the relationship between the structures of atoms and the properties of elements.

The number of protons in the nucleus of an atom defines the identity of that element. Since the number of electrons in a neutral atom is equal to the number of protons, an element's atomic number also reveals how many electrons are in that element's atoms. The electrons occupy regions of space that chemists and physicists call *shells.* The shells are further divided into regions of space called *subshells.* Subshells are related to angular momentum, which designates the shape of the electron orbit space around the nucleus. Shells are numbered 1, 2, 3, 4, and so forth (in theory out to infinity). In addition, shells may be designated by letters: The first shell is the K-shell, the second shell the L-shell, the third the M-shell, and so forth. Subshells have letter designations, s, p, d, and f being the most common. The *n*th shell has *n* possible subshells. Therefore, the first shell has only an s subshell, designated 1s; the second shell has both s and p subshells (2s and 2p); the third shell 3s, 3p, and 3d; and the fourth shell 4s, 4p, 4d, and 4f. (This pattern continues for higher-numbered shells, but this is enough for now.)

An s subshell is spherically symmetric and can hold a maximum of 2 electrons. A p subshell is dumbbell-shaped and holds 6 electrons, a d subshell 10 electrons, and an f subshell 14 electrons, with increasingly complicated shapes.

As the number of electrons in an atom increases, so does the number of shells occupied by electrons. In addition, because electrons are all negatively charged and tend to repel each other *electrostatically,* as the number of the shell increases, the size of the shell increases, which

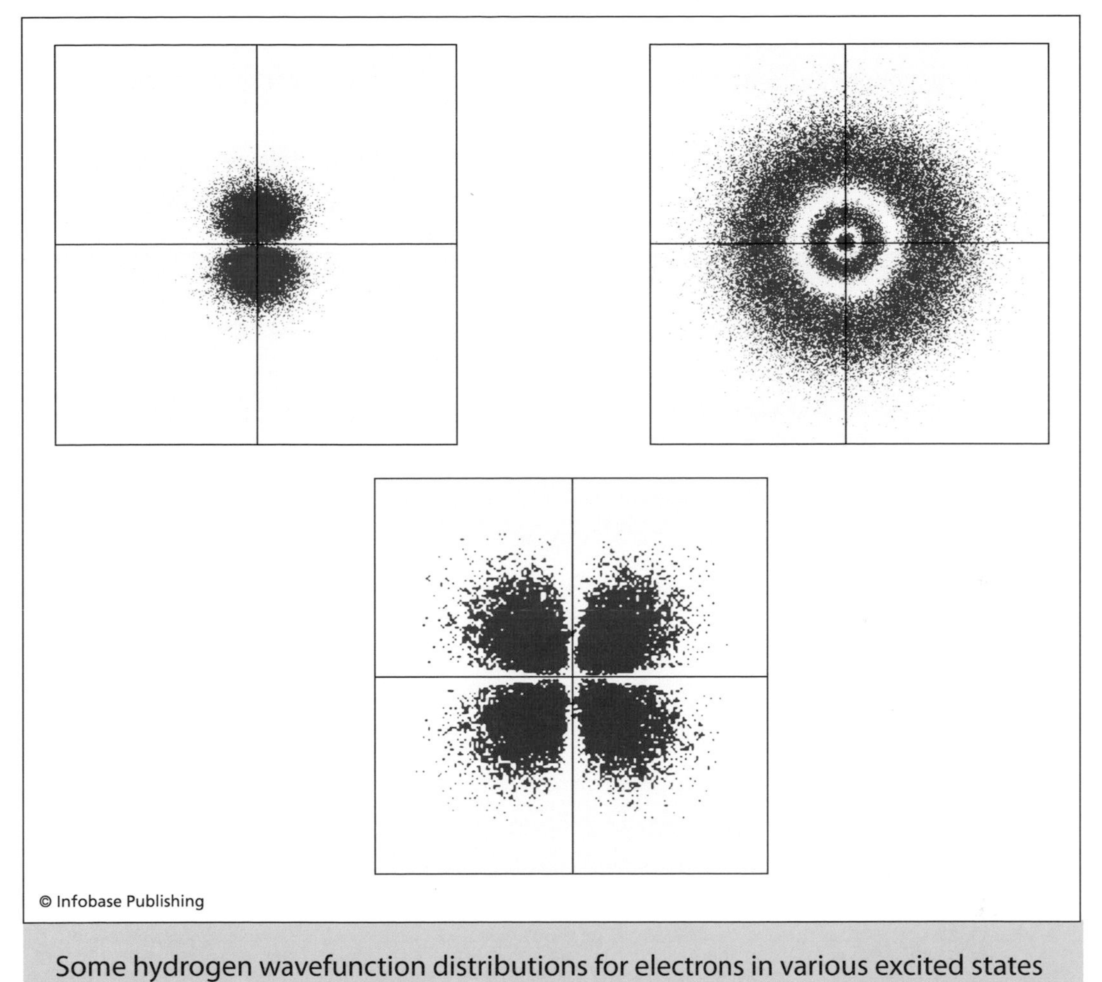

Some hydrogen wavefunction distributions for electrons in various excited states

means that electrons in higher-numbered shells are located, on the average, farther from the nucleus. Inner shells tend to be fully occupied with the maximum number of electrons they can hold. The electrons in the outermost shell, which is likely to be only partially occupied, will determine that atom's properties.

Physicists and chemists use *electronic configurations* to designate which subshells in an atom are occupied by electrons as well as how many electrons are in each subshell. For example, nitrogen is element number 7, so it has seven electrons. Nitrogen's electronic configuration is $1s^22s^22p^3$; a superscript designates the number of electrons that occupy a subshell. The first shell is fully occupied with its maximum of two electrons. The second shell can hold a maximum of eight electrons, but it is only partially occupied with just five electrons—two in the 2s subshell and three in the 2p. Those five outer electrons determine nitrogen's properties. For a heavy element like tin (Sn), electronic configurations can be quite complex. Tin's configuration is $1s^2 2s^2 2p^6 3s^2 3p^6 4s^2 3d^{10} 4p^6 5s^2 4d^{10} 5p^2$ but is more commonly written in the shorthand notation [Kr] $5s^2 4d^{10} 5p^2$, where [Kr] represents the electron configuration pattern for the noble gas krypton. (The pattern continues in this way for shells with higher numbers.) The important thing to notice about tin's configuration is that all of the shells except the last one are fully occupied. The fifth shell can hold 32 electrons, but in tin there are only four electrons in the fifth shell. The outer electrons determine an element's properties. The table on page xxix illustrates the electronic configurations for nitrogen and tin.

ATOMS ARE HELD TOGETHER WITH CHEMICAL BONDS

Fundamentally, a chemical bond involves either the sharing of two electrons or the transfer of one or more electrons to form ions. Two atoms of nonmetals tend to share pairs of electrons in what is called a *covalent bond*. By sharing electrons, the atoms remain more or less electrically neutral. However, when an atom of a metal approaches an atom of a nonmetal, the more likely event is the transfer of one or more electrons from the metal atom to the nonmetal atom. The metal atom becomes a positively charged ion and the nonmetal atom becomes a negatively charged ion. The attraction between opposite charges provides the force that holds the atoms together in what is called an *ionic bond*. Many chemical bonds are also intermediate in nature between covalent and ionic bonds and have characteristics of both types of bonds.

IN CHEMICAL REACTIONS, ATOMS REARRANGE TO FORM NEW COMPOUNDS

When a substance undergoes a *physical change*, the substance's name does not change. What may change is its temperature, its length, its *physical state* (whether it is a solid, liquid, or gas), or some other characteristic, but it is still the same substance. On the other hand, when a substance undergoes a *chemical change*, its name changes; it is a

ELECTRONIC CONFIGURATIONS FOR NITROGEN AND TIN

ELECTRONIC CONFIGURATION OF NITROGEN (7 ELECTRONS)				ELECTRONIC CONFIGURATION OF TIN (50 ELECTRONS)			
Energy Level	Shell	Subshell	Number of Electrons	Energy Level	Shell	Subshell	Number of Electrons
1	K	s	2	1	K	s	2
2	L	s	2	2	L	s	2
		p	3			p	6
			7	3	M	s	2
						p	6
						d	10
				4	N	s	2
						p	6
						d	10
				5	O	s	2
						p	2
							50

different substance. For example, water can decompose into hydrogen gas and oxygen gas, each of which has substantially different properties from water, even though water is composed of hydrogen and oxygen atoms.

In chemical reactions, the atoms themselves are not changed. Elements (like hydrogen and oxygen) may combine to form compounds (like water), or compounds can be decomposed into their elements. The atoms in compounds can be rearranged to form new compounds whose names and properties are different from the original compounds. Chemical reactions are indicated by writing chemical equations such as the equation showing the decomposition of water into hydrogen and oxygen: $2\ H_2O\ (l) \rightarrow 2\ H_2\ (g) + O_2\ (g)$. The arrow indicates the direction in which the reaction proceeds. The reaction begins with the *reactants* on the left and ends with the *products* on the right. We sometimes designate the physical state of a reactant or product in parentheses—*s* for solid, *l* for liquid, *g* for gas, and *aq* for *aqueous* solution (in other words, a solution in which water is the solvent).

IN NUCLEAR REACTIONS THE NUCLEI OF ATOMS CHANGE

In ordinary chemical reactions, chemical bonds in the reactant species are broken, the atoms rearrange, and new chemical bonds are formed in the product species. These changes only affect an atom's electrons; there is no change to the nucleus. Hence there is no change in an element's identity. On the other hand, nuclear reactions refer to changes in an atom's nucleus (whether or not there are electrons attached). In most nuclear reactions, the number of protons in the nucleus changes, which means that elements are changed, or transmuted, into different elements. There are several ways in which *transmutation* can occur. Some transmutations occur naturally, while others only occur artificially in nuclear reactors or particle accelerators.

The most familiar form of transmutation is *radioactive decay,* a natural process in which a nucleus emits a small particle or *photon* of light. Three common modes of decay are labeled *alpha, beta,* and *gamma* (the first three letters of the Greek alphabet). Alpha decay occurs among elements at the heavy end of the periodic table, basically elements heavier than lead. An alpha particle is a nucleus of helium 4 and is symbolized as $^{4}_{2}He$ or α. An example of alpha decay occurs when uranium 238 emits an alpha particle and is changed into thorium 234 as in the following reaction: $^{238}_{92}U \rightarrow ^{4}_{2}He + ^{234}_{90}Th$. Notice that the parent isotope, U-238, has 92 protons, while the daughter isotope, Th-234, has only 90 protons. The decrease in the number of protons means a change in the identity of the element. The mass number also decreases.

Any element in the periodic table can undergo beta decay. A beta particle is an electron, commonly symbolized as β^- or e^-. An example of beta decay is the conversion of cobalt 60 into nickel 60 by the following reaction: $^{60}_{27}Co \rightarrow ^{60}_{28}Ni + e^-$. The atomic number of the daughter isotope is one greater than that of the parent isotope, which maintains charge balance. The mass number, however, does not change.

In gamma decay, photons of light (symbolized by γ) are emitted. Gamma radiation is a high-energy form of light. Light carries neither mass nor charge, so the isotope undergoing decay does not change identity; it only changes its energy state.

Elements also are transmuted into other elements by nuclear *fission* and *fusion.* Fission is the breakup of very large nuclei (at least as heavy as uranium) into smaller nuclei, as in the fission of U-236 in the following reaction: $^{236}_{92}U \rightarrow ^{94}_{36}Kr + ^{139}_{56}Ba + 3n$, where n is the symbol for a neutron (charge = 0, mass number = +1). In fusion, nuclei combine to form larger nuclei, as in the fusion of hydrogen isotopes to make helium. Energy may also be released during both fission and fusion. These events may occur naturally—fusion is the process that powers the Sun and all other stars—or they may be made to occur artificially.

Elements can be transmuted artificially by bombarding heavy target nuclei with lighter projectile nuclei in reactors or accelerators. The transuranium elements have been produced that way. Curium, for example, can be made by bombarding plutonium with alpha particles. Because the projectile and target nuclei both carry positive charges, projectiles must be accelerated to velocities close to the speed of light to overcome the force of repulsion between them. The production of successively heavier nuclei requires more and more energy. Usually, only a few atoms at a time are produced.

ELEMENTS OCCUR WITH DIFFERENT RELATIVE ABUNDANCES

Hydrogen overwhelmingly is the most abundant element in the universe. Stars are composed mostly of hydrogen, followed by helium and only very small amounts of any other element. Relative abundances of elements can be expressed in parts per million, either by mass or by numbers of atoms.

On Earth, elements may be found in the lithosphere (the rocky, solid part of Earth), the hydrosphere (the aqueous, or watery, part of Earth), or the atmosphere. Elements such as the noble gases, the rare earths, and commercially valuable metals like silver and gold occur in only trace quantities. Others, like oxygen, silicon, aluminum, iron, calcium, sodium, hydrogen, sulfur, and carbon are abundant.

HOW NATURALLY OCCURRING ELEMENTS HAVE BEEN DISCOVERED

For the elements that occur on Earth, methods of discovery have been varied. Some elements—like copper, silver, gold, tin, and lead—have been known and used since ancient or even prehistoric times. The origins of their early metallurgy are unknown. Some elements, like phosphorus, were discovered during the Middle Ages by alchemists who recognized that some mineral had an unknown composition. Sometimes, as in the case of oxygen, the discovery was by accident. In other instances—as in the discoveries of the alkali metals, alkaline earths, and lanthanides—chemists had a fairly good idea of what they were looking for and were able to isolate and identify the elements quite deliberately.

To establish that a new element has been discovered, a sample of the element must be isolated in pure form and subjected to various chemical and physical tests. If the tests indicate properties unknown in any other element, it is a reasonable conclusion that a new element has been discovered. Sometimes there are hazards associated with isolating a substance whose properties are unknown. The new element could be toxic, or so reactive that it can explode, or extremely radioactive. During the course of history, attempts to isolate new elements or compounds have resulted in more than just a few deaths.

HOW NEW ELEMENTS ARE MADE

Some elements do not occur naturally, but can be synthesized. They can be produced in nuclear reactors, from collisions in particle accelerators, or can be part of the *fallout* from nuclear explosions. One of the elements most commonly made in nuclear reactors is technetium. Relatively large quantities are made every day for applications in nuclear medicine. Sometimes, the initial product made in an accelerator is a heavy element whose atoms have very short *half-lives* and undergo radioactive decay. When the atoms decay, atoms of elements lighter than the parent atoms are produced. By identifying the daughter atoms, scientists can work backward and correctly identify the parent atoms from which they came.

The major difficulty with synthesizing heavy elements is the number of protons in their nuclei ($Z > 92$). The large amount of positive charge makes the nuclei unstable so that they tend to disintegrate either by radioactive decay or *spontaneous fission*. Therefore, with the exception of a few transuranium elements like plutonium (Pu) and americium (Am), most artificial elements are made only a few atoms at a time and so far have no practical or commercial uses.

THE LANTHANIDES AND ACTINIDES SECTION OF THE PERIODIC TABLE

The elements in this book are found in the lower section of the periodic table and are called the lanthanides, actinides, and transactinides. They are introduced in groups as follows: lanthanides or rare earth elements; actinium, thorium, and protactinium; uranium; transuranium elements; and, transactinides. The key to understanding each element's information box that appears at the beginning of each chapter is on page xxxiv.

Included in the lanthanides and actinides are some of society's most industrially important substances. The lanthanide elements neodymium, samarium, and dysprosium are needed in the clean energy industry to make hybrid vehicles and wind turbines as efficient as possible. More than 50 percent of rare earth production goes into the steel industry. In nonferrous metal applications, rare earths are alloyed with aluminum and silicon to make pistons and with zinc to make corrosion-resistant coatings. Another element attracting interest in the energy sector is thorium. Thorium-fueled fission reactors could, in the long term, provide a large fraction of the energy demand of the planet more efficiently and with fewer hazardous side effects than current-day uranium enrichment reactors. Uranium-fueled reactors will continue to provide power, but the waste factor for both uranium and plutonium is a crucial issue.

In this book, readers will learn about the important properties of the lanthanides and actinides and how these elements are useful and possibly hazardous in everyday life.

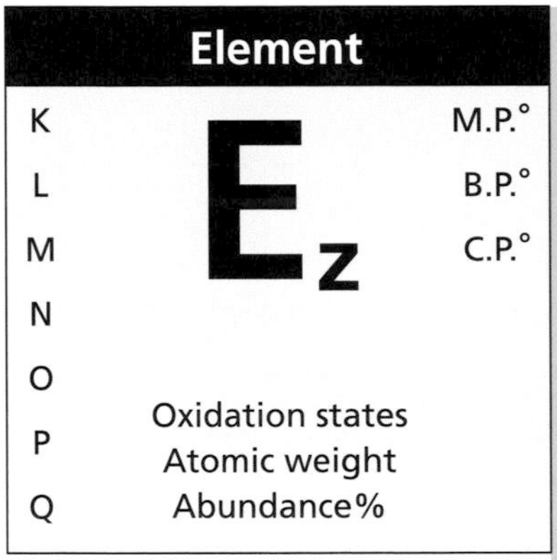

Element		
K		M.P.°
L	E_Z	B.P.°
M		C.P.°
N		
O		
P	Oxidation states Atomic weight	
Q	Abundance%	

Information box key. E represents the element's letter notation (H=Hydrogen), with the Z subscript indicating proton number. Orbital shell notations appear in the column on the left. For elements that are not naturally abundant, the mass number of the longest-lived isotope is given in brackets. The abundances (atomic %) are based on meteorite and solar wind data. The melting point (M.P.), boiling point (B.P.), and critical point (C.P.) temperatures are given in Celsius. Should sublimation and critical point temperatures apply, these are indicated by s and t, respectively.

1

The Lanthanides or Rare Earth Elements

Lanthanum (element 57) is included in this book in the discussion of the lanthanides. Otherwise, the substances referred to as the *lanthanides* will be considered to consist of the 14 elements with atomic numbers 58 to 71: cerium (Ce, 58), praseodymium (Pr, 59), neodymium (Nd, 60), promethium (Pm, 61), samarium (Sm, 62), europium (Eu, 63), gadolinium (Gd, 64), terbium (Tb, 65), dysprosium (Dy, 66), holmium (Ho, 67), erbium (Er, 68), thulium (Tm, 69), ytterbium (Tb, 70), and lutetium (Lu, 71). The term *lanthanide* suggests that all of these elements have chemical and physical properties remarkably similar to those of the element lanthanum. Indeed, the similarities are so strong that these elements are considered to occupy a single place in the periodic table directly under the element yttrium (Y, number 39). Listing them separately in the table requires their being split out into a separate row that normally is shown below the body of the periodic table itself. Because of the manner in which they appear to be grouped *inside* the transition metals, one name for the lanthanides is *inner transition metals.*

Another name for the lanthanides is *rare earth elements* (or *rare earths*), which dates to the time when the abundances of these elements on Earth were thought to be extremely low. Some chemists also use the term *rare earths* to include 17 elements—scandium, yttrium, lanthanum, and the 14 lanthanides listed above. In the present discussion, however, scandium and yttrium will not be treated.

Today, geologists know that the lanthanides are more abundant than had previously been believed. In fact, several of the lanthanides are more abundant than the platinum metals even though elements such as platinum are likely more familiar to the general public. The difference is that metals such as copper and tin are found in concentrated deposits, unlike the rare earths, which tend to be widely dispersed. It should be noted that, as a general rule, where one rare earth is found, several other rare earths will also be found in the same ores, an association that results from their similar chemical properties.

The most common oxidation state of lanthanide elements in aqueous solution is "+3," and it is in the "+3" state that these elements are most commonly found in ores. The three principal ores of the lanthanides are monazite, a phosphate; bastnasite, a carbonate; and xenotime, also a phosphate. Monazite and bastnasite mostly contain lanthanum, cerium, praseodymium, and neodymium. Xenotime mostly contains cerium, neodymium, samarium, gadolinium, dysprosium, erbium, ytterbium, and yttrium.

The "4f" electrons in the lanthanides are referred to as *inner electrons,* meaning that they lie closer to the nucleus than "5s" and "5p" electrons do. Because the "4f" electrons are effectively *shielded* from the electrons of other atoms to which lanthanide atoms might be bonded, the "4f" electrons for the most part do not participate in chemical bonding.

In this chapter, the reader will learn about the discoveries of the rare earth elements, the chemistry of these elements, and their applications in modern society.

(continues on page 10)

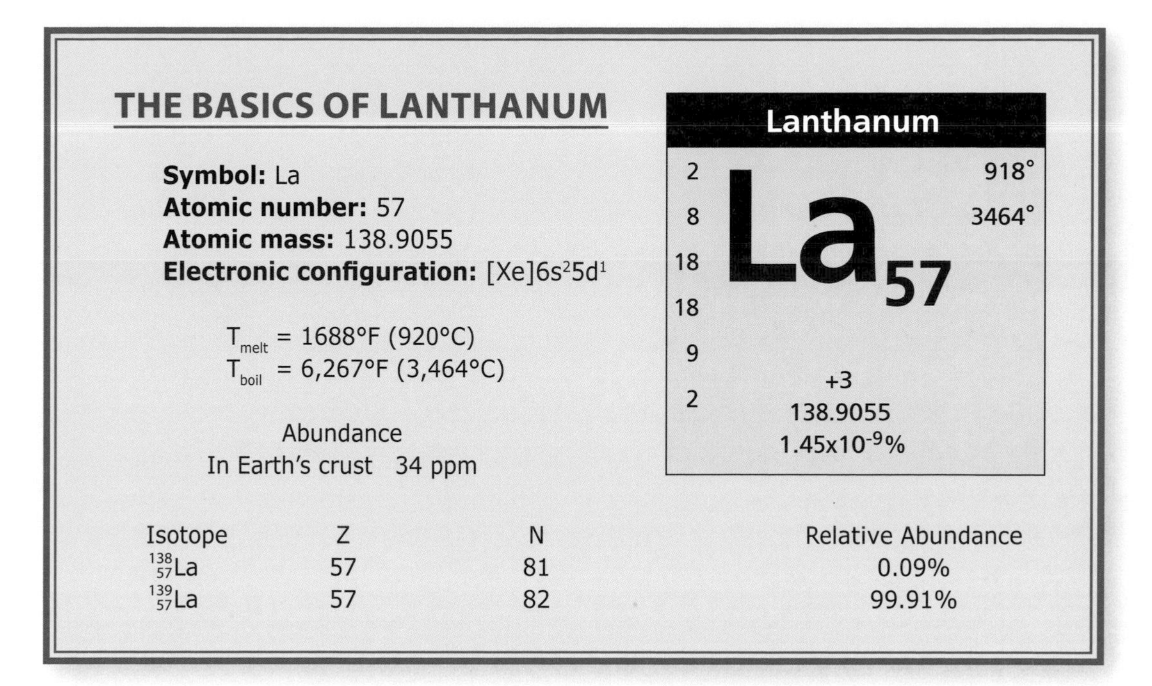

THE BASICS OF LANTHANUM

Symbol: La
Atomic number: 57
Atomic mass: 138.9055
Electronic configuration: $[Xe]6s^2 5d^1$

T_{melt} = 1688°F (920°C)
T_{boil} = 6,267°F (3,464°C)

Abundance
In Earth's crust 34 ppm

Lanthanum
2 8 18 18 9 2
La 57
918°
3464°
+3
138.9055
1.45×10^{-9}%

Isotope	Z	N	Relative Abundance
$^{138}_{57}La$	57	81	0.09%
$^{139}_{57}La$	57	82	99.91%

THE BASICS OF CERIUM

Symbol: Ce
Atomic number: 58
Atomic mass: 140.116
Electronic configuration:
$[Xe]6s^2 4f^1 5d^1$

T_{melt} = 1,470°F (799°C)
T_{boil} = 6,229°F (3,443°C)

Abundance
In Earth's crust 60 ppm

Cerium
2 8 18 19 9 2
Ce 58
799°
3443°
+3+4
140.116
3.70×10^{-9}%

Isotope	Z	N	Relative Abundance
$^{136}_{58}Ce$	58	78	0.185%
$^{138}_{58}Ce$	58	80	0.251%
$^{140}_{58}Ce$	58	82	88.450%
$^{142}_{58}Ce$	58	84	11.114%

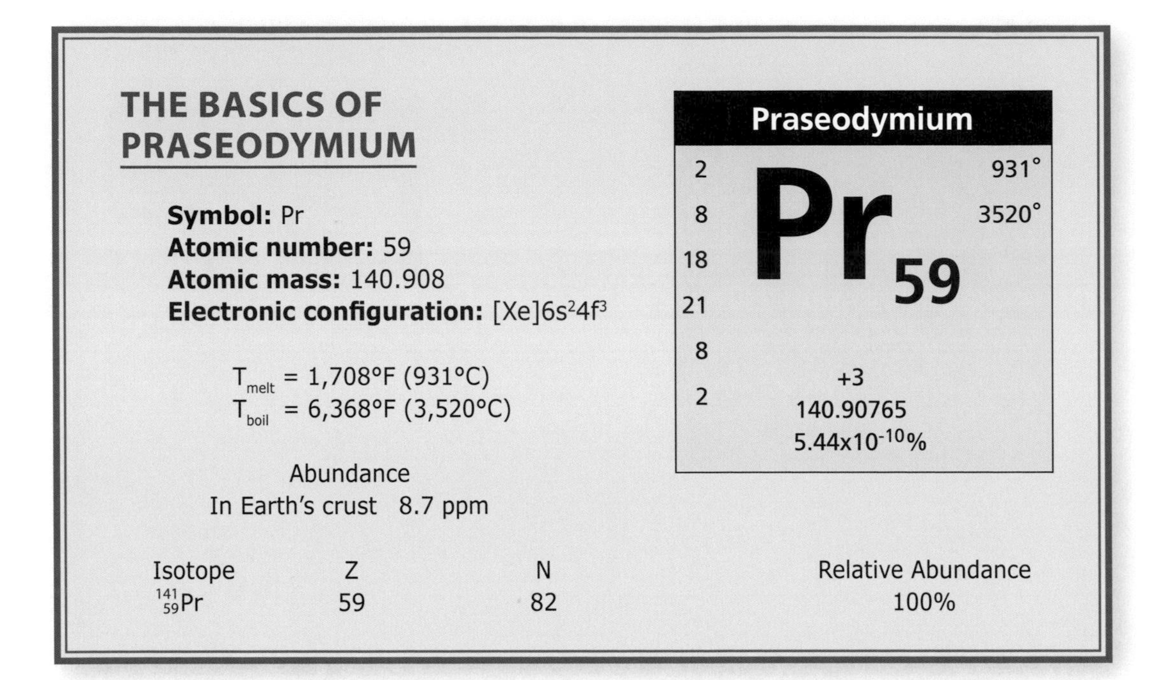

THE BASICS OF PRASEODYMIUM

Symbol: Pr
Atomic number: 59
Atomic mass: 140.908
Electronic configuration: $[Xe]6s^24f^3$

T_{melt} = 1,708°F (931°C)
T_{boil} = 6,368°F (3,520°C)

Abundance
In Earth's crust 8.7 ppm

Isotope	Z	N	Relative Abundance
$^{141}_{59}Pr$	59	82	100%

THE BASICS OF NEODYMIUM

Symbol: Nd
Atomic number: 60
Atomic mass: 144.24
Electronic configuration: $[Xe]6s^24f^4$

T_{melt} = 1,861°F (1,016°C)
T_{boil} = 5,565°F (3,074°C)

Abundance
In Earth's crust 33 ppm

Neodymium
2
8
18
22
8
2
Nd
60
1021°
3074°
+3
144.24
$2.70 \times 10^{-9}\%$

Isotope	Z	N	Relative Abundance
$^{142}_{60}Nd$	60	82	27.2%
$^{143}_{60}Nd$	60	83	12.2%
$^{144}_{60}Nd$	60	84	23.8%
$^{145}_{60}Nd$	60	85	8.3%
$^{146}_{60}Nd$	60	86	17.2%
$^{148}_{60}Nd$	60	88	5.7%
$^{150}_{60}Nd$	60	90	5.6%

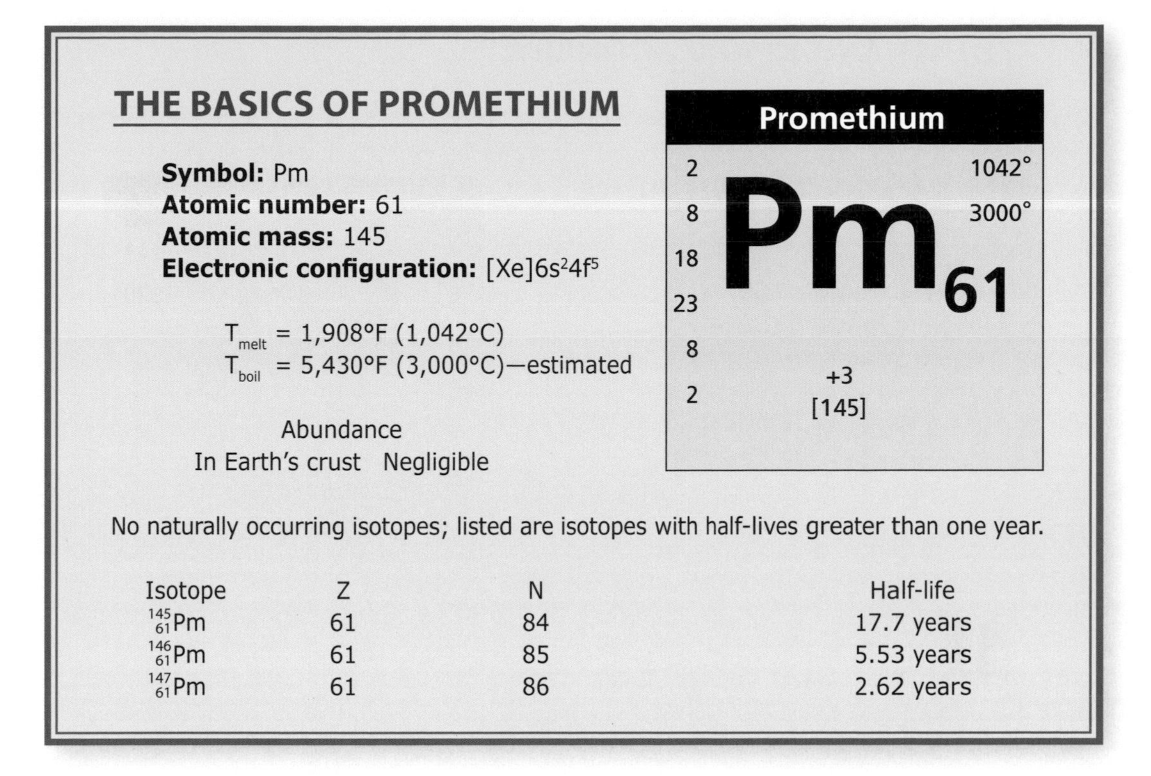

THE BASICS OF PROMETHIUM

Symbol: Pm
Atomic number: 61
Atomic mass: 145
Electronic configuration: [Xe]$6s^2 4f^5$

T_{melt} = 1,908°F (1,042°C)
T_{boil} = 5,430°F (3,000°C)—estimated

Abundance
In Earth's crust Negligible

Promethium
2
8
18
23
8
2
1042°
3000°
Pm
61
+3
[145]

No naturally occurring isotopes; listed are isotopes with half-lives greater than one year.

Isotope	Z	N	Half-life
$^{145}_{61}$Pm	61	84	17.7 years
$^{146}_{61}$Pm	61	85	5.53 years
$^{147}_{61}$Pm	61	86	2.62 years

THE BASICS OF SAMARIUM

Symbol: Sm
Atomic number: 62
Atomic mass: 150.36
Electronic configuration: [Xe]$6s^2 4f^6$

T_{melt} = 1,962°F (1,072°C)
T_{boil} = 3,261°F (1,794°C)

Abundance
In Earth's crust 6 ppm

Samarium
2
8
18
24
8
2
1074°
1794°
Sm
62
+2+3
150.36
8.42×10^{-10}%

Isotope	Z	N	Relative Abundance
$^{144}_{62}$Sm	62	82	3.07%
$^{147}_{62}$Sm	62	85	14.99%
$^{148}_{62}$Sm	62	86	11.24%
$^{149}_{62}$Sm	62	87	13.82%
$^{150}_{62}$Sm	62	88	7.38%
$^{152}_{62}$Sm	62	90	26.75%
$^{154}_{62}$Sm	62	92	22.75%

THE BASICS OF EUROPIUM

Symbol: Eu
Atomic number: 63
Atomic mass: 151.964
Electronic configuration: $[Xe]6s^2 4f^7$

T_{melt} = 1,513°F (822°C)
T_{boil} = 2,905°F (1,596°C)

Abundance
In Earth's crust 1.8 ppm

Isotope	Z	N	Relative Abundance
$^{151}_{63}Eu$	63	88	47.81%
$^{153}_{63}Eu$	63	90	52.19%

Europium

2		822°
8		1596°
18	Eu_{63}	
25		
8	+2+3	
2	151.964	
	3.17×10^{-10}%	

THE BASICS OF GADOLINIUM

Symbol: Gd
Atomic number: 64
Atomic mass: 157.25
Electronic configuration:
$[Xe]6s^2 4f^7 5d^1$

T_{melt} = 2,395°F (1,313°C)
T_{boil} = 5,923°F (3,273°C)

Abundance
In Earth's crust 5.2 ppm

Isotope	Z	N	Relative Abundance
$^{152}_{64}Gd$	64	88	0.20 %
$^{154}_{64}Gd$	64	90	2.18%
$^{155}_{64}Gd$	64	91	14.80%
$^{156}_{64}Gd$	64	92	20.47%
$^{157}_{64}Gd$	64	93	15.65%
$^{158}_{64}Gd$	64	94	24.84%
$^{160}_{64}Gd$	64	96	21.86%

Gadolinium

2		1313°
8		3273°
18	Gd_{64}	
25		
9	+3	
2	157.25	
	1.076×10^{-9}%	

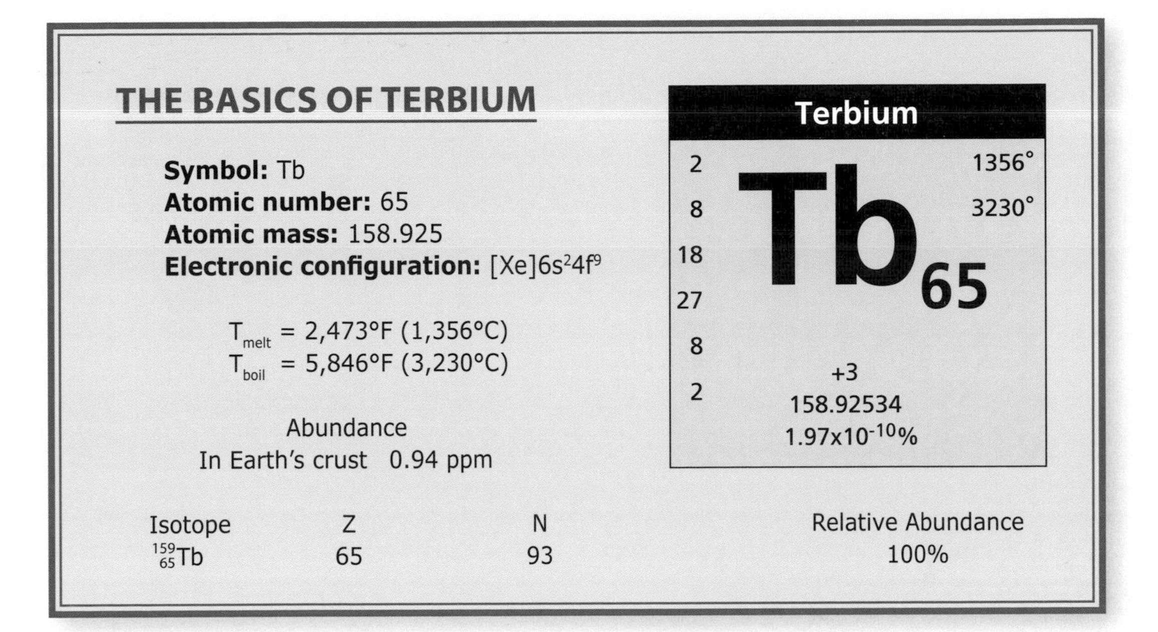

THE BASICS OF TERBIUM

Symbol: Tb
Atomic number: 65
Atomic mass: 158.925
Electronic configuration: [Xe]$6s^2 4f^9$

T_{melt} = 2,473°F (1,356°C)
T_{boil} = 5,846°F (3,230°C)

Abundance
In Earth's crust 0.94 ppm

Terbium
2
8
18
27
8
2
Tb 65
1356°
3230°
+3
158.92534
1.97×10^{-10}%

Isotope	Z	N	Relative Abundance
$^{159}_{65}Tb$	65	93	100%

THE BASICS OF DYSPROSIUM

Symbol: Dy
Atomic number: 66
Atomic mass: 162.50
Electronic configuration: [Xe]$6s^2 4f^{10}$

T_{melt} = 2,574°F (1,412°C)
T_{boil} = 4,653°F (2,567°C)

Abundance
In Earth's crust 6.2 ppm

Dysprosium
2
8
18
28
8
2
Dy 66
1412°
2567°
+3
162.50
1.286×10^{-9}%

Isotope	Z	N	Relative Abundance
$^{156}_{66}Dy$	66	90	0.06%
$^{158}_{66}Dy$	66	92	0.10%
$^{160}_{66}Dy$	66	94	2.340%
$^{161}_{66}Dy$	66	95	18.91%
$^{162}_{66}Dy$	66	96	25.51%
$^{163}_{66}Dy$	66	97	24.90%
$^{164}_{66}Dy$	66	98	28.18%

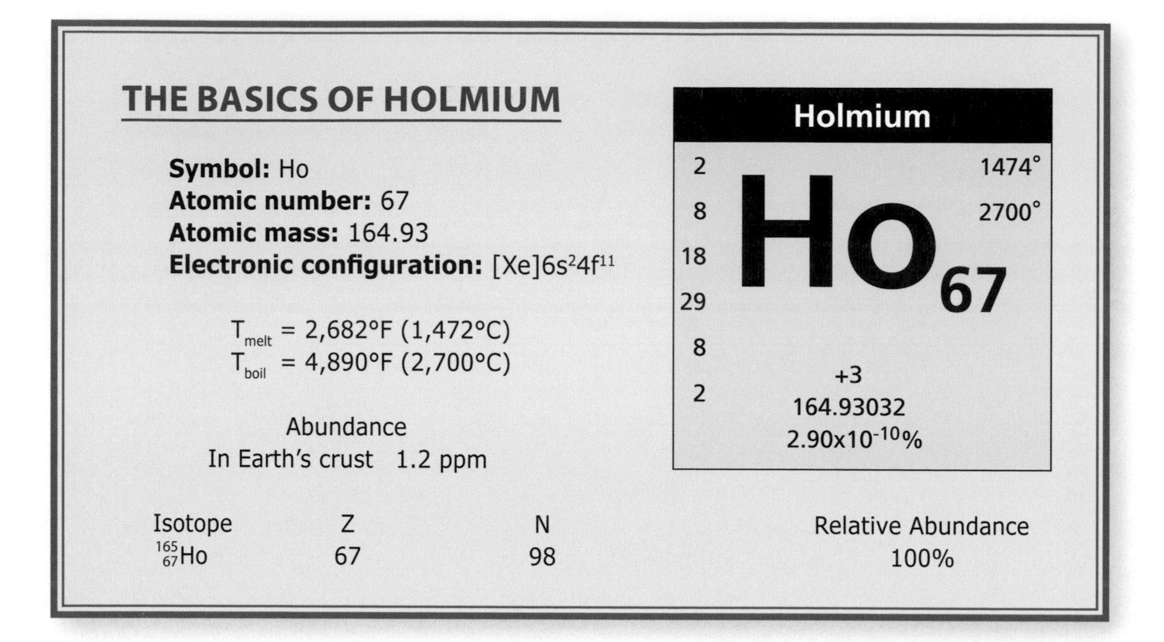

THE BASICS OF HOLMIUM

Symbol: Ho
Atomic number: 67
Atomic mass: 164.93
Electronic configuration: $[Xe]6s^2 4f^{11}$

T_{melt} = 2,682°F (1,472°C)
T_{boil} = 4,890°F (2,700°C)

Abundance
In Earth's crust 1.2 ppm

Isotope	Z	N	Relative Abundance
$^{165}_{67}Ho$	67	98	100%

THE BASICS OF ERBIUM

Symbol: Er
Atomic number: 68
Atomic mass: 167.26
Electronic configuration: $[Xe]6s^2 4f^{12}$

T_{melt} = 2,874°F (1,529°C)
T_{boil} = 5,194°F (2,868°C)

Abundance
In Earth's crust 3.0 ppm

Erbium
2
8
18
30
8
2
1529°
2868°
Er
68
+3
167.26
8.18×10^{-10}%

Isotope	Z	N	Relative Abundance
$^{162}_{68}Er$	68	94	0.14%
$^{164}_{68}Er$	68	96	1.61%
$^{166}_{68}Er$	68	98	33.61%
$^{167}_{68}Er$	68	99	22.93%
$^{168}_{68}Er$	68	100	26.78%
$^{170}_{68}Er$	68	102	14.93%

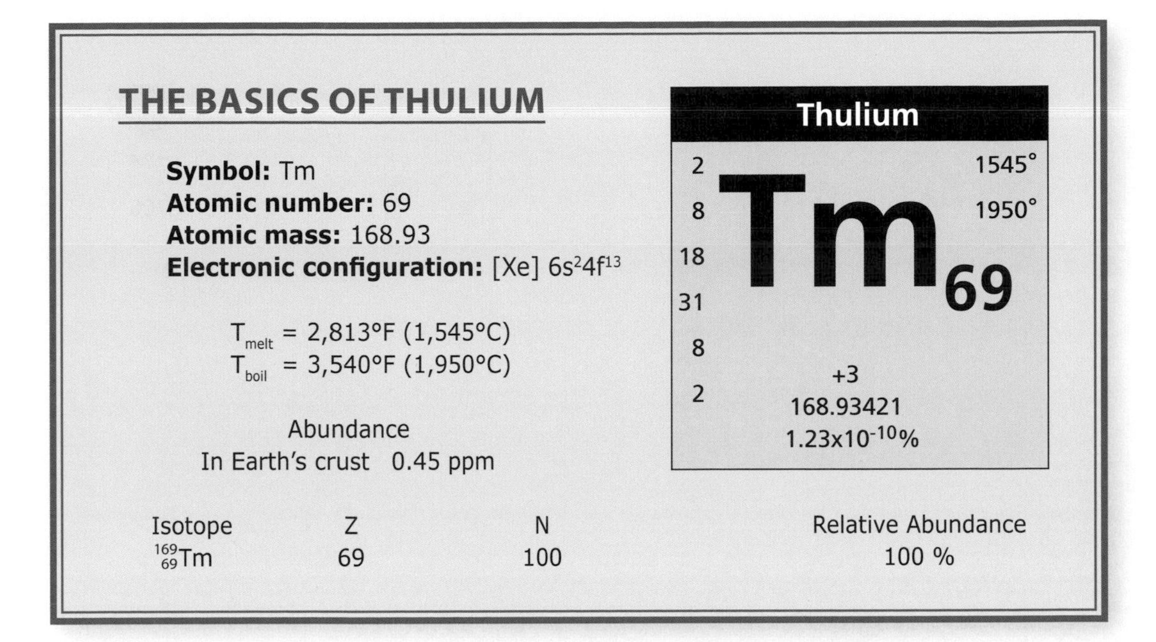

THE BASICS OF THULIUM

Symbol: Tm
Atomic number: 69
Atomic mass: 168.93
Electronic configuration: [Xe] $6s^2 4f^{13}$

T_{melt} = 2,813°F (1,545°C)
T_{boil} = 3,540°F (1,950°C)

Abundance
In Earth's crust 0.45 ppm

Isotope	Z	N	Relative Abundance
$^{169}_{69}$Tm	69	100	100 %

Thulium
2
8
18
31
8
2
1545°
1950°
Tm
69
+3
168.93421
1.23×10^{-10}%

THE BASICS OF YTTERBIUM

Symbol: Yb
Atomic number: 70
Atomic mass: 173.04
Electronic configuration: [Xe]$6s^2 4f^{14}$

T_{melt} = 1,515°F (824°C)
T_{boil} = 2,185°F (1,196°C)

Abundance
In Earth's crust 2.8 ppm

Isotope	Z	N	Relative Abundance
$^{168}_{70}$Yb	70	98	0.13%
$^{170}_{70}$Yb	70	100	3.04%
$^{171}_{70}$Yb	70	101	14.28%
$^{172}_{70}$Yb	70	102	21.83%
$^{173}_{70}$Yb	70	103	16.13%
$^{174}_{70}$Yb	70	104	31.83%
$^{176}_{70}$Yb	70	105	12.76%

Ytterbium
2
8
18
32
8
2
819°
1196°
Yb
70
+2+3
173.04
8.08×10^{-10}%

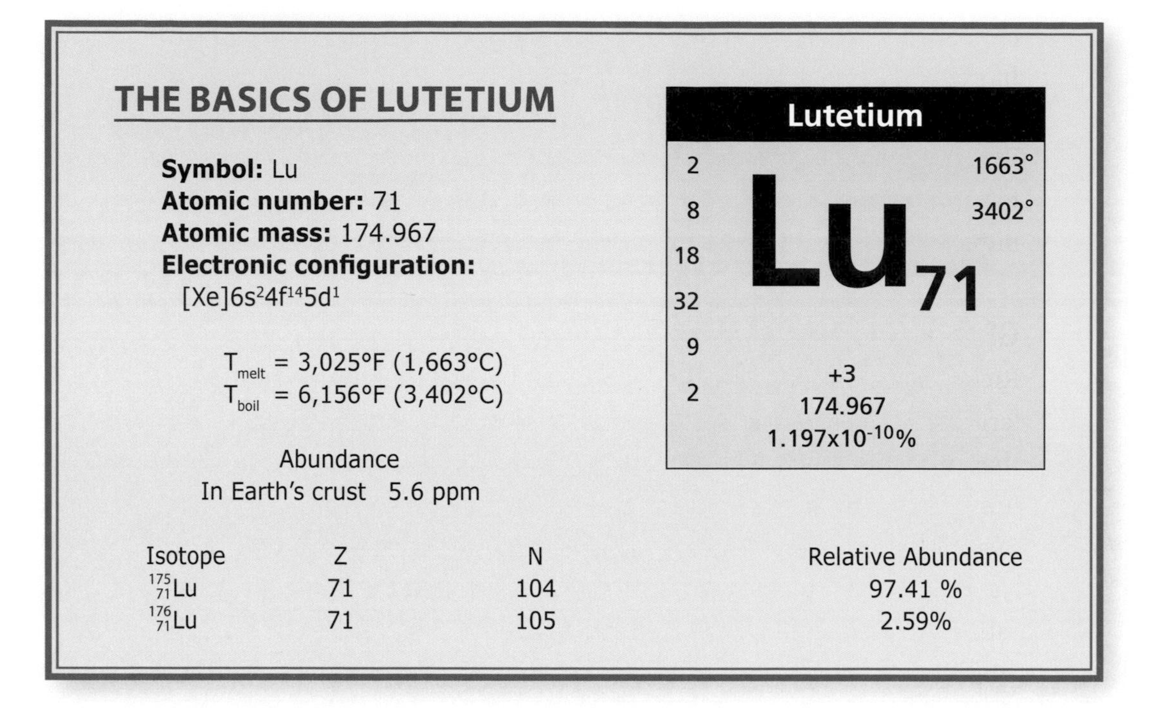

THE BASICS OF LUTETIUM

Symbol: Lu
Atomic number: 71
Atomic mass: 174.967
Electronic configuration:
$[Xe]6s^2 4f^{14} 5d^1$

T_{melt} = 3,025°F (1,663°C)
T_{boil} = 6,156°F (3,402°C)

Abundance
In Earth's crust 5.6 ppm

Isotope	Z	N	Relative Abundance
$^{175}_{71}Lu$	71	104	97.41 %
$^{176}_{71}Lu$	71	105	2.59%

(continued from page 2)

THE ASTROPHYSICS OF THE LANTHANIDES

Most of the lanthanide elements are synthesized in *supernova* explosions via the so-called *r-process,* where free, rapidly moving neutrons collide with and are captured by other nuclei. Relatively few stars exhibit spectral lines that identify lanthanides in their atmospheres, however, because these elements are heavy and tend to sink into the star's core rather than remaining near the surface, where their excitation spectra could be observed.

Spectral lines of several lanthanide ions, including lanthanum, cerium, neodymium, samarium, europium, gadolinium, and praseodymium, have been noted in high-mass *chemically peculiar* (CP) *stars* of various types. In general, the hotter CP stars with strong magnetic fields show an excess (above solar abundance) of lanthanide elements. In such stars, the slow neutron process *(s-process)* can contribute to lanthanide abundance. In this process, nuclei with masses greater than or equal to that of iron absorb relatively slow neutrons over long periods of time to form heavier elements. On the other hand, some cooler CP stars appear to be completely missing neodymium and samarium from their spectra. In many of these peculiar stars, strange magnetic field distributions are involved along with an unknown mechanism for separating isotopes. Because the physics of CP stars is so complex, this area of research has fostered the development of novel instrumentation and modeling techniques and will probably continue to do so for years to come.

GEOLOGY: THE CALIFORNIA DEPOSIT

Rich veins of ore containing lanthanides are extremely rare. The first large deposit of the mineral bastnasite—which contains cerium, lanthanum, and neodymium, along with small amounts of praseodymium, europium, samarium, and gadolinium—was discovered in the late 1940s by prospectors seeking uranium in the mountains of California. At the time, no useful purpose was known for these elements, so it is rather surprising that one of the prospectors was able to convince his employer (Molybdenum Corporation of America) to stake a mining claim on the land, which became known as the Mountain Pass Mine. Within 15 years, however, demand had skyrocketed for europium oxide, which was needed for increased production of color televisions. At the same time, cigarette lighters, which use cerium in the flints, saw ballooning sales. By the mid-1960s, the mine was producing more than 20 million pounds of rare earth concentrates per year. Soon thereafter, microwave ovens became popular, requiring supplies of samarium for the control mechanism. Neodymium, praseodymium, and gadolinium were also in demand for use in glass science. Production ramped up, and by 1990 the Mountain Pass Mine was the world's leading producer of rare earth elements, providing about 40 percent of global needs.

Separation of the desirable from undesirable constituents of bastnasite is, however, a messy process. The ore is dug from an open pit and ground into powder, after which a flotation procedure is used: The heavier rare-earth-containing portions fall to the bottom, while the lighter materials wash away. Unfortunately, radioactive by-products are also washed away, and the company became subject to legal action due to contamination by spills. In 1998, as a result of pressure by environmental groups and competition from a rare earth mine in China, the company suspended its mining operations.

In the mid-1960s, the California Mountain Pass Mine produced more than 20 million pounds (9.1 million kg) of rare earth concentrates per year. *(USGS Photographic Collection)*

Since 1998, the Bayan Obo mine (also known as Bajun Obo or Bayun Obo) in China has been providing 90 percent of the world's supply of rare earth metal oxides (see graph on page 13). Reports that Chinese exports of these metals will be drastically curtailed in the near future has prompted a reopening of the Mountain Pass Mine with enhanced safeguards against environmental contamination. Molycorp Minerals, a private investor group that purchased the facility in 2008, is currently processing stockpiled bastnasite concentrate and plans to restart mining of fresh ore in 2011.

DISCOVERY AND NAMING OF THE RARE EARTHS

The discoveries and subsequent separations and isolations of the rare earth elements were hindered by the elements' scarcity and by their very similar chemistries. In most cases, the identifications of the ele-

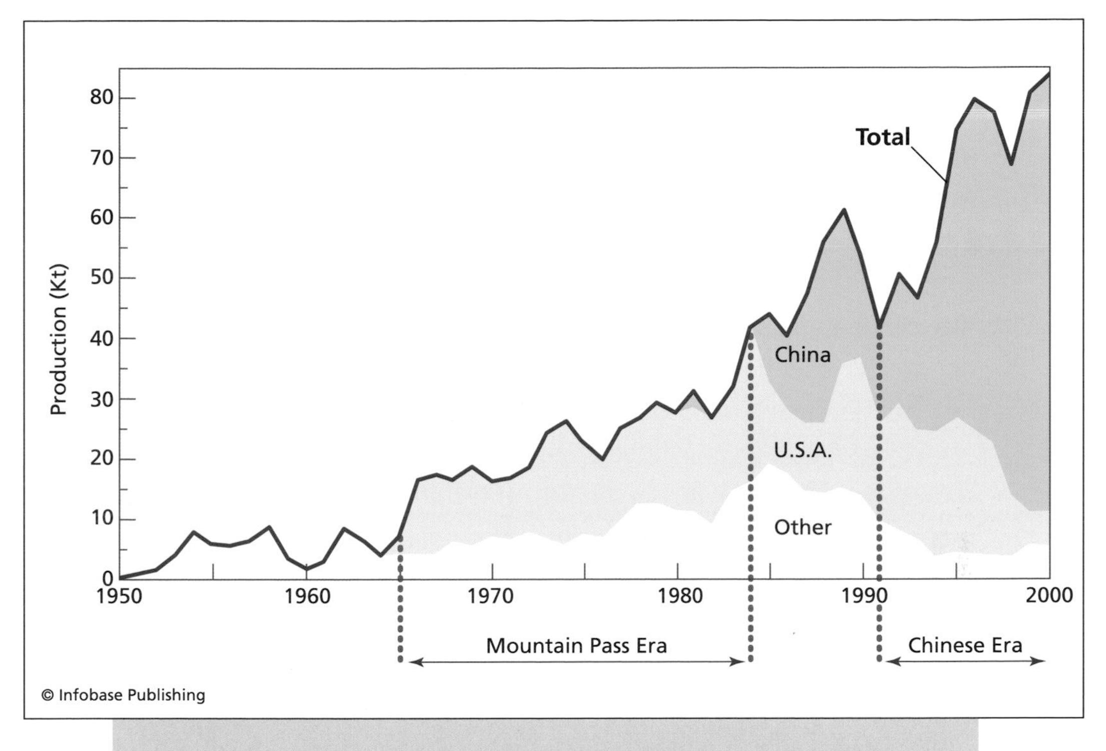

This graph of global production of rare earth oxides shows the global production of rare earth oxides in kilotons for the years 1950–2000. The Mountain Pass Mine suspended mining in 1998.

ments were made using oxides ("earths") of the elements, not the pure elements themselves. Only much later were pure samples of the elements finally obtained. One of the elements—promethium—has no stable isotopes, so its discovery waited until the 20th century, when it could be made artificially. The remaining rare earths had to be painstakingly separated from one another, a process that required more than 100 years to complete.

Before giving the history, some explanation of nomenclature is necessary for what follows. First of all, the term *earth* was used originally for what today chemists would identify as the oxide of an element. (The term *rare earth* dates from before the discovery of oxygen and reflects the perceived scarcity of these substances.) The rare earths occur together in a relatively small number of ores. Because their chemistry is so extremely similar, separations of rare earths are much more difficult than separations of elements in other parts of the periodic table. In addition, in the 1700s and 1800s, analytical techniques were fairly unsophisticated by today's standards. As a consequence, chemists often were not completely sure of what it was that they had produced. Many times, it was unclear to them whether they had actually isolated a pure element or still had a mixture of elements, or perhaps a pure compound, or maybe even a mixture of compounds. Since most of the rare earths were discovered before the concept of atomic number was developed, chemists were hindered by not even knowing how many rare earth elements should fit into the periodic table.

Physical and chemical techniques for separating or purifying substances matured only during the 19th and 20th centuries. An important tool used to identify the composition of substances was emission spectroscopy, which was developed by Robert Bunsen (1811–99) and Gustav Kirchhoff (1824–87). In emission spectroscopy, a small sample of an element or compound is heated to a temperature sufficient to cause the substance to glow. Each element emits a characteristic pattern of lines (called a *spectrum*) that is unique to that element. The pattern of lines observed tells chemists what elements are in the sample. Additionally, it was not until well into the 20th century that chemists were able to obtain samples of high purity of most metals. The result is that keeping track of the ores, compounds, and pure elements pertaining to the rare earths can be confusing at times. As a general rule, the names of ores or minerals, which can be mixtures of a large number of substances, tend to end with the suffix *-ite.* Oxides of rare earths (which are compounds, not pure elements) that are found in the ores end with the suffix *-ia.* The elements themselves end with the suffix *-ium.* The nomenclature can be confusing sometimes because 19th-century chemists themselves were often confused!

The story of the rare earths begins in two quarries in Sweden. In 1751, the Swedish chemist and mining expert Axel Fredrik Cronstedt (1722–65) discovered in Vestmanland a reddish-brown mineral that was originally named *tungsten,* or "heavy stone," which simply reflects its heaviness. The name was later changed to *cerite,* which is the name that will be used here. (The term *tungsten* today refers to a completely different element.) Lanthanum and the rare earths cerium, praseodynmium, neodymium, samarium, gadolinium, and europium were all eventually isolated from cerite, although the discoveries took more than 100 years to occur. [See the following flow chart.]

In 1787, the Swedish army lieutenant Carl Axel Arrhenius (1757–1824) picked up a black rock near the town of Ytterby near Stockholm. Arrhenius named the rock *ytterbite* after the town. In 1794, the Finnish chemist Johan Gadolin (1760–1852) analyzed the rock, which he renamed *ytterite,* and decided that it contained a new element which was called *yttrium* in honor of its place of discovery. Still later, the ore was again renamed—this time *gadolinite* in honor of Gadolin. (*Gadolinium* is the name still in use today.) Scandium and the rare earths erbium, terbium, ytterbium, holmium, thulium, dysprosium, and lutetium were eventually isolated from gadolinium, also over a 100-year period. [See the flow chart on page 16.] The events leading to the

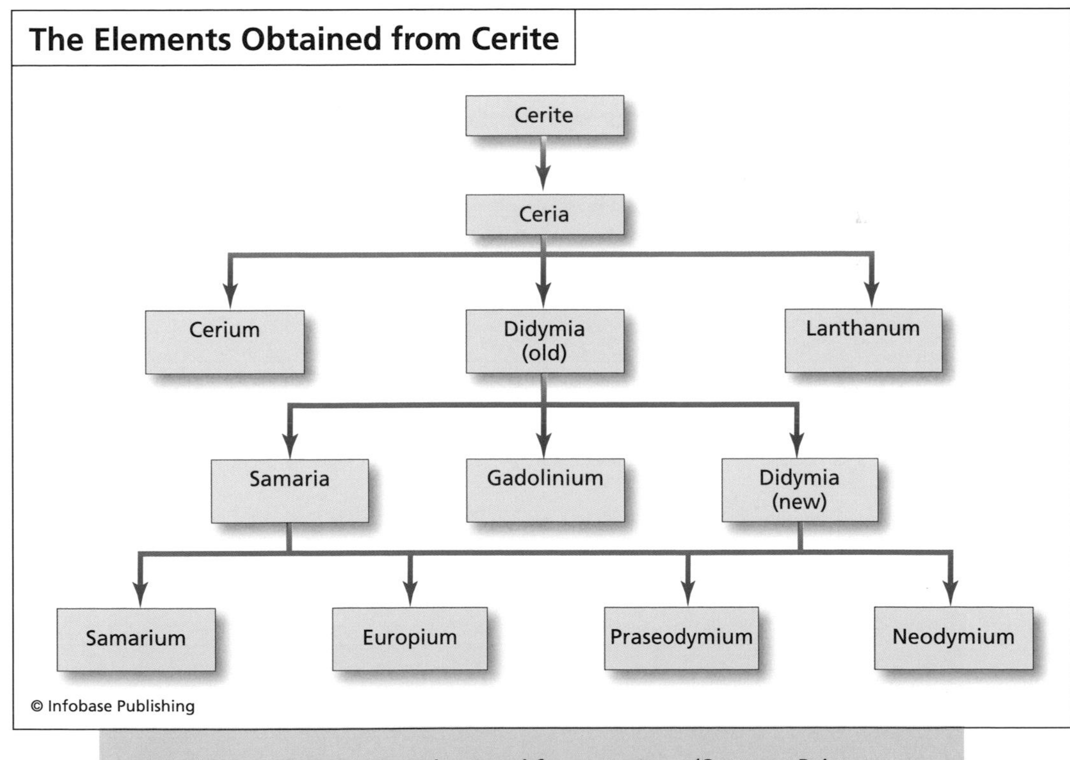

Flow chart of elements obtained from cerite. *(Source: Brian Nordstrom)*

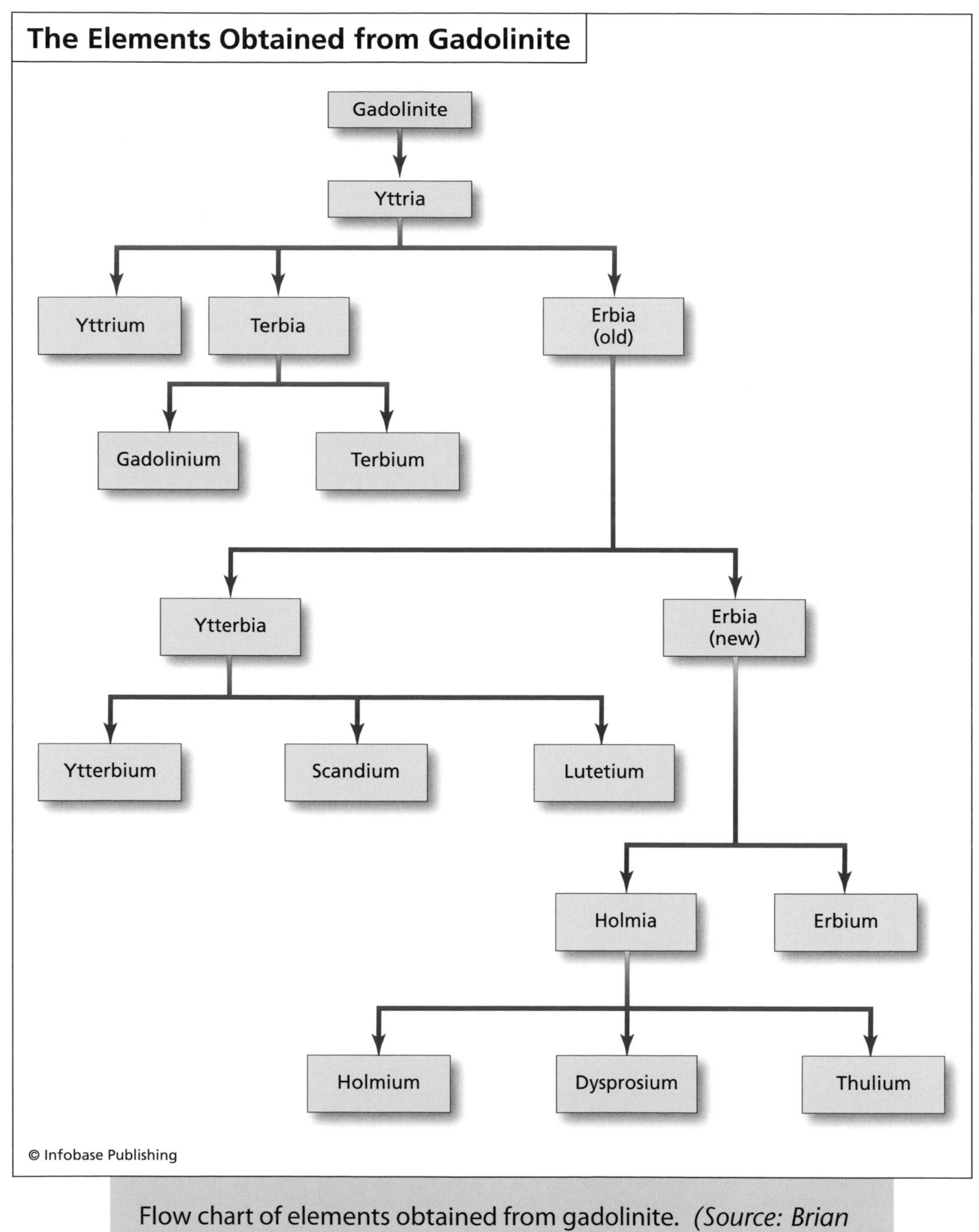

Flow chart of elements obtained from gadolinite. *(Source: Brian Nordstrom)*

discoveries of these two sets of elements are related below. After 1907, only promethium remained to be discovered. Because promethium has no stable isotopes, it does not occur in nature. It had to be synthesized artificially, an achievement that finally occurred in 1945. Promethium's late addition to the periodic table made it the last element lighter than uranium to be discovered.

Because Gadolin devoted much of his life to investigating the properties of the rare earth elements, his name is associated more with the rare earths than is the name of any other chemist. In Gadolin's honor, element 64 was named *gadolinium.*

In 1803, the Swedish chemists Jöns Jacob Berzelius (1799–1848) and Wilhelm Hisinger (1766–1852) investigated cerite and announced the discovery of an oxide that was named *ceria.* Independently, the German chemist Martin Klaproth (1743–1817) made the same discovery. The name *ceria* was chosen because the asteroid Ceres had just been discovered just two years earlier. Element 58 (cerium) was eventually isolated from ceria, although it required another 72 years to accomplish that feat. A pure sample of cerium was finally obtained in 1875 in Washington, D.C., by the American chemists William Francis Hillebrand (1853–1925) and Thomas H. Norton (1851–1941), who worked at the National Bureau of Standards (today, the National Institute of Standards and Technology, or NIST).

The Swedish chemist Jöns Jacob Berzelius was an important figure in the isolation of elements from cerite. *(Dibner Library of the History of Science and Technology)*

JOHAN GADOLIN

Johan Gadolin was born in Åbo, Finland, on June 5, 1760. At that time, Finland was part of Sweden. His father was a professor of physics and theology at Åbo University. Johan entered the university at the young age of 14. Under the inspiration of Professor Pehr Gadd (1727–97), Johan chose to major in chemistry. At the age of 19, Johan transferred to the University of Uppsala in Sweden, where he studied chemistry under Professor Torbern Bergman (1735–85) and eventually graduated.

Gadolin became an accomplished linguist who could read and write a variety of languages, including Swedish, Finnish, English, French, German, Russian, and Latin. Gadolin traveled extensively throughout Europe, visiting leading universities and meeting many prominent scientists. In 1789, he returned to Åbo University, where he joined the chemistry faculty and became a full professor upon the death of Pehr Gadd.

At the age of 34, Gadolin married Hedvig Tihleman. Together they had nine children. When Gadolin was 56 years old, Hedvig died. Three years later, he married the 24-year-old Ebba Palander. Gadolin spent little time with his family, however, being passionately devoted to his work.

Gadolin is remembered in part for his scientific papers on specific heats of gases, minerals, and metals, but also in part for being an outstanding teacher who influenced an entire generation of students. In 1798, he published the first modern Swedish chemistry textbook. Gadolin's major fame rests on his comprehensive study of the rare earth elements. At the time of Gadolin's birth, there was no periodic table nor even an unambiguous definition of what an element is. Many of the elements known today had not yet been discovered, including such common substances as hydrogen, nitrogen, and oxygen. The discoveries of the rare earths resulted largely from Gadolin's work.

Gadolin's investigation of the rare earths began with the discovery in 1787 of a black ore that was named *ytterbite* because it was found in a feldspar quarry at Ytterby, a town near Stockholm, Sweden. (Feldspar was used in the manufacture of porcelain.) In

Gadolinium was named after the Finnish chemist Johan Gadolin. *(A. B. F. Tilgmann)*

1794, Gadolin obtained a sample of ytterbite, which he renamed *ytterite* and from which he isolated an oxide, yttria, which was later shown to contain the element yttrium. The name *ytterbite* was later changed to *gadolinite* in Gadolin's honor. Almost a century later, the element gadolinium was isolated from gadolinite and named for Gadolin. Altogether, during his lifetime, Gadolin investigated the compositions of the minerals that ultimately yielded all of the rare earths with the exception of promethium (since promethium does not occur naturally). Gadolin retired from Åbo University in 1822.

On August 15, 1852, Johan Gadolin died in Wirmo, Finland, at the age of 92. At the time of his death, his personal library numbered more than 3,600 books, most of which still exist. Two of his laboratories are also still in existence, although they now serve other purposes. It is most appropriate that his name is remembered in the element named after him—gadolinium, element 64.

In 1839, the Swedish chemist Carl Gustav Mosander (1797–1858) conducted his own investigation of ceria. (Remarkably, Mosander was already beginning to develop cataracts by this time and lived the last few years of his life almost totally blind.) Starting with the compound cerium nitrate, Mosander was able to isolate a new element that he named *lanthanum* (element 57), which comes from a Greek word meaning "hidden" because it was hidden among compounds of cerium. Two years later, Mosander isolated a reddish-colored oxide from lanthanum that he named *didymium,* meaning "twin brother" of lanthanum. Didymium found use in glass goggles used to protect the eyes of glassblowers. However, it turned out that didymium was not in fact a pure element, but a mixture of five elements—samarium, gadolinium, europium, praseodymium, and neodymium. To think that cerium oxide—believed at one time to be a pure substance—eventually yielded seven different elements!

Suspicion regarding the purity of didymium began in 1853 with the Swiss chemist Jean-Charles Galissard de Marignac (1817–94). In 1879, the French chemist Paul-Émile Lecoq de Boisbaudran (1838–1912) iso-

Metallic europium *(Visuals Unlimited)*

lated a new element from didymium oxide that he named *samarium* (element 62) after the mineral samarskite in which samarium can be found. (Samarskite itself was named in honor of the Russian mining official Vasili Samarsky-Bykhovets [1803–70]. Although indirectly, it could be claimed that Samarsky-Bykhovets was actually the first person to have an element named for him.)

In 1880, Marignac isolated a new oxide from samarskite. Boisbaudran suggested the new compound be named *gadolinia,* and Marignac agreed. By that time, the rare earths had become closely associated with Gadolin's name, and in 1886, when a new element was found in gadolinia, it was named *gadolinium* (element 64) in Gadolin's memory. Thus, Johan Gadolin is generally credited as the first person for whom an element was named.

By this time, Dmitri Mendeleev had published his periodic table. Because the concept of atomic number had not yet been developed, Mendeleev arranged the elements in order of atomic weight. That posed a problem for the placement of the rare earths in the table since their atomic weights were all found to be intermediate between the weights of lanthanum and hafnium, which according to Mendeleev's way of arranging elements should be adjacent to each other. Normally, the difference in atomic weights between two consecutive elements is only about one to five grams per mole. The difference between lanthanum and hafnium, however, is almost 40 grams per mole. Clearly, there was room for a number of elements to fit between lanthanum and hafnium, but according to Mendeleev's table, there was no room for new elements. The analogs of lanthanum and hafnium are the elements yttrium and zirconium, which are located in the row above lanthanum and hafnium. Since there were no known elements lying between them, there appeared to be no room for any new elements to lie between lanthanum and hafnium.

This dilemma was not resolved until 1913 (after Mendeleev's death), when the English physicist Henry Moseley (1887–1915) proposed the concept of atomic number. According to Moseley, elements should be arranged in order of nuclear charge (i.e., atomic number) instead of atomic weight. By using data obtained from the rare earths' X-ray spectra, Moseley assigned atomic numbers of 57 to lanthanum and 72 to hafnium. Accordingly, Moseley predicted that there was room for 14 new elements to lie between them. These 14 elements proved to be the rare earths. Unfortunately for Moseley, his life was cut short soon afterward. In 1915, he was killed in Gallipoli, Turkey, during World War I.

In 1882, the Czech chemist Bohuslav Brauner (1855–1935) observed two sets of unidentified spectral lines in a sample of didymium. Three years later, the Austrian chemist Baron Auer von Welsbach (1858–1929)—a student of Robert Bunsen's—isolated two new rare earth elements, which he named *praseodynium* (element 59), which means "green twin," and *neodymium* (element 60), which means "new twin," both names in reference to the original sample of didymium. In 1902, Brauner predicted that an element should exist that would be located between neodymium and samarium in the periodic table. After Brauner's death, that element was produced artificially and named *promethium.*

In 1901, another new rare earth element was isolated. The French chemist Eugène-Anatole Demarçay (1852–1904) observed unidentified spectral lines in a sample of samarium magnesium nitrate. Demarçay named the new substance *europium* (element 63) after the continent of Europe.

To recap the story to this point, the reader is reminded that all of these events resulted from the collection in 1751 of the reddish-brown mineral cerite by Axel Cronstedt in a Swedish quarry. The discoveries of lanthanum and the rare earths cerium, praseodynmium, neodymium, samarium, gadolinium, and europium all resulted from Cronstedt's original find. The story now returns to Carl Arrhenius, who, in 1787, collected an interesting-looking black rock in a quarry near Ytterby, Sweden, and who took the rock home for further study. This rock, ytterbite, or ytterite (later gadolinite), was eventually found to contain the element scandium and the rare earths erbium, terbium, ytterbium, holmium, thulium, dysprosium, and lutetium.

In 1843, after completing his investigations of ceria, Carl Mosander turned his attention to the study of yttria. His experiments resulted in the discoveries of two new rare earth compounds—terbia (eventually shown to contain terbium, element 65) and erbia (eventually shown to contain erbium, element 68). Both elements were named for the town of

Ytterby. (Actually, Mosander's original assignment of the oxide names *terbia* and *erbia* were later reversed. The element names *terbium* and *erbium* were assigned after the oxide names had been reversed, so that the nomenclature used here reflects the names in use today.)

In 1878, Jean-Charles Marignac conducted an investigation of erbia and found a new rare earth compound in it, which he named *ytterbia.* Ytterbia was later shown to contain ytterbium (element 70), which was also named for the town of Ytterby in Sweden. In 1879, the Swedish chemist Lars Nilson (1840–99), working in Uppsala, Sweden, further investigated ytterbia and obtained from it the new element scandium (number 21), which was named in honor of Scandinavia. Nilson demonstrated that scandium satisfied the properties of ekaboron (an element that would be in the same family as boron) that had been predicted by Mendeleev.

In Geneva, Switzerland, in 1878, the chemists Marc Delafontaine (1837–1911) and Jacques-Louis Soret (1827–90) studied the spectral lines emitted by heated erbia and observed lines that could not be attributed to any known element. At the same time, in Uppsala, Sweden, the chemist Per Cleve (1840–1905) investigated the composition of erbia and demonstrated that it had three components that he named *erbia, holmia,* and *thulia.* Cleve showed that holmia contained a new rare earth element that he named *holmium* (element 67) after the city of Stockholm. He demonstrated also that holmium was the element whose spectrum had been detected previously by Delafontaine and Soret in Switzerland. It was not until 1911, however, that pure holmium was finally isolated. Delafontaine, Soret, and Cleve together are credited as the discoverers of holmium.

The rare earth element thulium (number 69) was isolated from Cleve's compound thulia. The name *thulium* was derived from Thule, an ancient name for Scandinavia. Like holmium, pure thulium was not obtained until 1911.

In 1886, Paul-Émile Lecoq de Boisbaudran further decomposed holmium and found another compound in it that he named *dysprosia.* Dysprosia was later shown to contain the rare earth element dysprosium (number 66), its name being derived from Greek and meaning "hard to obtain." Dysprosium was found to be one of the more common rare earth elements, exceeding in relative abundance more familiar elements such as boron, bromine, and mercury.

In 1905, Auer von Welsbach separated ytterbia into two separate compounds. In 1906, he isolated two new elements from these compounds—element number 70, which he named *aldebaranium,* and element 71, which he named *cassiopeium,* both names being derived from the Greek names for the star Aldebaran and the constellation Cassiopeia. Auer procrastinated in publishing his findings, however, an error that cost him recognition for priority of discovery and that resulted in these elements being given names different from the ones he had proposed.

In 1907, before Auer had published his findings, the French chemist Georges Urbain (1872–1938) independently investigated the composition of ytterbia and also concluded that it contained two substances—one for which he retained the name *ytterbia* and the other that he named *lutecia,* a named derived from an old name for Paris. The latter name was later changed to *lutetia,* from which the name of the element lutetium (number 71) was derived. At the same time that Urbain was doing his work, the English-American chemist Charles James (1880–1926) of the University of New Hampshire, who was unaware of the discoveries of either of the other two men, also discovered lutetium. In addition to this specific discovery, James was probably the best-known chemist of the early 20th century in terms of his very thorough investigations of rare earth chemistry, the separation of the rare earth elements, and their purification.

A serious dispute arose between Auer and Urbain over priority of discovery. Today, Urbain and James are usually jointly credited with lutetium's discovery. Generally, Auer's contribution is overlooked. In the early 1900s, however, Auer and Urbain were both so highly esteemed by the chemical community for their investigations of rare earth chemistry that both men received numerous nominations for Nobel Prizes. That neither man was ever awarded a Nobel Prize may have reflected the reluctance of the Nobel committee to became involved in the controversy between them.

As of 1907, only promethium was still missing from the row of lanthanide elements in the periodic table. For the 13 lanthanides that had been discovered at that time, known deposits were still mostly found in Sweden. That situation shifted, however, with the discovery of substantial deposits in India, Brazil, South Africa, and the Mohave Desert of Southern California. Today, China leads the world in rare earth element production.

The discovery of promethium followed an entirely different track than the discoveries of the other rare earth elements because promethium had to be produced artificially. As a result of Moseley's work, chemists realized that an unknown element had to exist with an atomic number of 61, which would place it between neodymium and samarium in the periodic table. Various claims were made in the 1920s by investigators who thought they had isolated element 61 from natural sources. None of these claims, however, were validated by other researchers. Indeed, promethium cannot be found in purely natural sources because no natural processes produce it, and its longest-lived isotope (Pm-145) has a half-life of only 17.7 years.

The discovery of promethium occurred in 1945 at the Clinton Laboratories at Oak Ridge National Laboratory in Tennessee. The American chemists Charles Coryell (1912–71), Lawrence Glendenin (1918–), and Jacob Marinsky (1918–) found isotopes of all of the rare earth elements among the fission products of uranium. They were able to isolate small quantities of element 61, which they originally proposed to name *clintonium* after their laboratory. Coryell's wife, Grace, however, suggested the name *prometheum* after the Greek mythological god Prometheus, who stole fire from the gods and gave it to humans. The name *prometheum* was accepted but changed to *promethium* to make it uniform with the names of other elements that end in *-ium*. The discovery of promethium closed the book on finding any more chemical elements lighter than uranium.

THE LANTHANIDE CONTRACTION

Elements in the periodic table exhibit general trends in the sizes of their atoms. Descending a column of the periodic table, atoms tend to become larger in size because their valence electrons occupy successively higher energy levels. The result is to place the outermost electrons farther from the nucleus. On the other hand, proceeding across a row (or period) of the table from left to right, atoms tend to become smaller in size. Each increase in atomic number results in the addition of one proton to an atom's nucleus and one electron to the region outside the nucleus. Since the electron added to the next element is usually in the same energy level as the outermost electrons of the element preceding it, there is some tendency for the atom to increase in size just because of the additional repulsion that has been added to the electrons. However, that tendency is more than counterbalanced by the effect of the increase in positive charge on the nucleus. The increased charge on the nucleus exerts an attractive force that tends to pull all of the atom's electrons closer to the nucleus. Thus, the result of moving from left to right across a row is a gradual decrease (or *contraction*) in the sizes of atoms.

This contraction in atomic size is especially pronounced in the case of the lanthanides. It is important enough to be given its own name—the *lanthanide contraction.* From lanthanum (element 57) to lutetium (element 71), a total of 14 electrons are successively added to the elements' "4f" subshells. The result is a decrease of about 20 percent in atomic size from the beginning of the series to the end. There at least two reasons for the greater decrease in size in the "4f" series than in any of the "s," "p," or "d" series. First, an "s" series has only two elements, a "p" series only six elements, and a "d" series only 10 elements, whereas an "f" series has 14 elements. Given a steady decrease in size in moving from one element to the next, it would be expected that the cumulative effect would be greater with a larger number of elements. Second, the orbitals in the different types of subshells are shaped differently. In "s," "p," and "d" orbitals, the inner, or *core,* electrons can partially *shield* the outer electrons from the attractive force exerted by the nucleus. In the case of "f" orbitals, however, the shapes of the orbitals cause the electron density to be spread out more diffusely through the atom, resulting in a lesser shielding effect. Therefore, the nuclei can exert stronger pulls on the outermost electrons, drawing them closer to the nucleus and giving the atoms smaller sizes.

There are two important consequences of the lanthanide contraction. The first consequence is the grouping of the lanthanides into two subgroups: lanthanum, cerium, praseodymium, neodymium, and samarium (and promethium, although it does not occur naturally), which together are referred to as the *ceria subgroup;* and europium, gadolinium, terbium, dysprosium, holmium, erbium, thulium, ytterbium, and lutetium, which together are referred to as the *yttria subgroup.* The first group's elements have larger atoms than those of the second group. Consequently, the elements making up the ceria subgroup have chemical properties more similar to one another than they do to elements in the second group. What is true of lanthanum, which is located in the third row of the transition metals, or cerium, is equally true of the others in that group. On the other hand, the members of the yttria subgroup have atoms that are significantly smaller and similar in size to the element yttrium, which is located in the second row of the transition metals above lanthanum. Consequently, the properties of the elements in the yttria subgroup more closely resemble the properties of yttrium than they do the properties of the elements located earlier in the lanthanide series.

The second consequence of the lanthanide contraction is its effect on the elements in the third row of the transition metals. Elements in the first row of transition metals are characterized by electrons in the "3d" subshell. Elements in the second row are characterized by electrons in the "4d" subshell. Following the general trend of increasing atomic size upon descending a column of the table, the atoms of "4d" metals are significantly larger in size than the atoms of metals located just above them—the "3d" metals, and the chemical and physical properties of "4d" metals are significantly different from the properties of "3d" metals. However, it is not true that atoms of "5d" metals are significantly larger in size than atoms of "4d" metals. Because of the lanthanide contraction, by the time lutetium is reached, the atoms have decreased to the size of the first "4d" metal (yttrium), again with the result that the size of atoms of each "5d" metal is almost identical to the size of atoms of the "4d" metal located just above it. The result is that the chemical and physical properties of "4d" and "5d" transition metals are very similar to one another, and correspondingly very different from the properties of "3d" metals. In other words, in the titanium group, for example, zirconium and hafnium are virtually indistinguishable from each other, but are both very different from titanium. Likewise, in the nickel group, palladium and platinum have extremely similar properties, but are very different from nickel. The same comparison can be made in each of the groups, or families, of transition metals.

THE CHEMISTRY OF THE LANTHANIDES

All of the pure rare earth metals are difficult to prepare. Attempts at reducing lanthanide salts with strong reducing agents such as sodium or magnesium usually yield alloys of the lanthanides rather than the pure metals themselves. Consequently, the more frequent method of preparation is to use electrolytic techniques. In an electrolytic cell, an electrical current is passed through an ionic solution. There are two electrodes—an *anode* and a *cathode.* Metal ions are reduced to their pure metallic state and collected at the cathode.

Once prepared, the elements are all gray, soft, malleable, and ductile. They tend to be very reactive metals, comparable to magnesium. Lanthanum metal, for example, burns brightly in air according to the following chemical reaction:

$$4\ La\ (s) + 3\ O_2\ (g) \rightarrow 2\ La_2O_3\ (s).$$

The metals react vigorously with both strong and weak acids. For example, the reaction of neodymium with hydrochloric acid is shown in the following equation:

$$2\ Nd\ (s) + 6\ HCl\ (aq) \rightarrow 2\ NdCl_3\ (aq) + 3\ H_2\ (g).$$

At room temperature, the metals react slowly with water to form insoluble hydroxides. Europium is the most reactive of the lanthanides with water as shown by the following reaction:

$$Eu\ (s) + 3\ H_2O\ (l) \rightarrow Eu(OH)_2 \cdot H_2O\ (s) + H_2\ (g).$$

(Note that europium is in the "+2" oxidation state in the product; "+2" is an unusual oxidation state for a lanthanide.)

The historical discoveries of the lanthanide elements resulted from the discoveries of the ores cerite and gadolinite (yttrite), containing ceria and yttria, respectively. Both ores contain at least tiny amounts of all of the lanthanides, but ceria generally has a much higher content of lanthanum, cerium, praseodymium, neodymium, and samarium, which together are referred to as the ceria subgroup. (Since promethium does not occur naturally, it is omitted from this group.) Yttria generally contains more europium, gadolinium, terbium, dysprosium, holmium, erbium, thulium, ytterbium, and lutetium, which together are referred to as the *yttria subgroup.* The division into these subgroups results from the lanthanide contraction (described above). Elements in the ceria subgroup have larger atoms, and elements in the yttria subgroup have smaller atoms. Lanthanum atoms are similar in size to the atoms of elements in the ceria subgroup, giving it similar chemistry. Yttrium atoms are similar in size to the atoms of the yttria subgroup, giving it similar chemistry.

The chemistry of the lanthanide elements is dominated by the "+3" oxidation state, which is the primary oxidation state in aqueous solutions. In the solid state, there are a few exceptions, namely cerium, praseodymium, and terbium, which exhibit a "+4" oxidation state, and europium, ytterbium, and samarium, which exhibit a "+2" state.

The chemistry of these elements is so similar that is very difficult to isolate them from one another using chemical techniques. In the 19th and early 20th centuries, the only method that was available was *fractional crystallization,* which makes use of slightly small differences in the solubilities of the elements' salts. The result is that the two elements are not completely separated; just their ratio changes. As two elements co-precipitate, a slightly greater amount of one of the elements is left in solution. *Decanting,* or *filtering,* the solution gives a solution with a slightly greater concentration of one of the elements. Conversely, the precipitate becomes more concentrated in the other element. Repeated precipitations eventually yield reasonably pure samples of the elements. To illustrate, when the first separation of ytterbium and lutetium occurred in 1907, the chemists performed 15,000 fractional crystallizations! The separations of lanthanides were so difficult that, had it not been for absorption and emission spectroscopy, it would have been virtually impossible to judge how complete the separations were.

During the 20th century, the much more efficient technique of *ion-exchange chromatography* was developed. In ion-exchange chromatography, a solution containing ions of two or more elements is passed through a column containing a resin. As the solution passes through, the various ions to be separated are absorbed onto the resin. Another solution, an acid for example, called the *eluant* is poured through. The ions that were adhering to the column are removed at different rates and, thus, pass out of the column at different times. As each ion leaves the column, it can be collected in a beaker or flask. The term *chromatography* refers to the common visual aid to the separation that the ions may be different colors.

All of the trivalent ions form compounds with common anions. Thus, there are compounds of hydroxide (OH^-), carbonate (CO_3^{2-}), sulfate (SO_4^{2-}), nitrate (NO_3^-), phosphate (PO_4^{3-}), oxide (O^{2-}), oxalate ($C_2O_4^{2-}$), and the halides (F^-, Cl^-, Br^-, I^-). The nitrates, chlorides, bromides, and iodides are water-soluble. The fluorides, hydroxides, oxides, carbonates, phosphates, and oxalates are insoluble. The solubilities of the sulfates vary.

Cerium forms a "+4" ion, Ce^{4+}. The solid oxide, CeO_2, is similar to plumbic oxide, PbO_2, in that both are oxidizing agents. For example, in aqueous solution Ce^{4+} oxidizes iodide (I^-) to molecular iodine (I_2) as shown in the following equation:

$$2\,Ce^{4+}\,(aq) + 2\,I^-(aq) \rightarrow 2\,Ce^{3+}\,(aq) + I_2\,(aq).$$

Cerium nitrate forms a soluble orange-red solid with ammonium nitrate. The solid has the formula $Ce(NO_3)_4 \cdot 2\,NH_4NO_3$.

Although samarium, europium, and ytterbium can form compounds in the "+2" oxidation state, this condition is regarded as anomalous. In aqueous solution, Sm^{2+}, Eu^{2+}, and Yb^{2+} are all strong reducing agents, as they themselves are easily oxidizing to the "+3" ions: Using Eu^{2+} as an example, these three ions will reduce I_2 to I^- as shown in the following reaction:

$$2\,Eu^{2+}\,(aq) + I_2\,(aq) \rightarrow 2\,Eu^{3+}\,(aq) + 2\,I^-(aq).$$

The properties of Ce^{4+} as an oxidizing agent and of Eu^{2+} as a reducing agent suggest that these two ions will also react with each other, as demonstrated in the following reaction:

$$Ce^{4+}\,(aq) + Eu^{2+}\,(aq) \rightarrow Ce^{3+}\,(aq) + Eu^{3+}\,(aq).$$

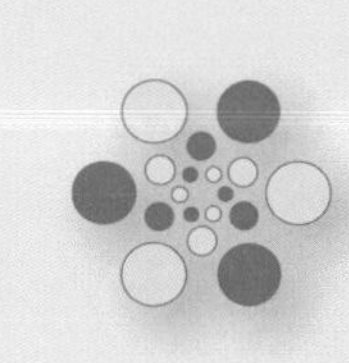

CLEAN ENERGY AND THE RARE EARTH ELEMENTS

A vital reason for reinstating the mining of rare earth elements at the Mountain Pass mine in California is their importance in clean energy technology. In order to develop energy sources that do not emit carbon dioxide, magnetic materials that have good temperature stability and *magnetic flux* retention are key. Neodymium and samarium are the best elements to supply these properties when alloyed with other metals. Permanent magnets made of neodymium iron boron (NdFeB)—often doped with dysprosium to improve certain magnetic behavior—and samarium cobalt (SmCo) can replace the more cumbersome *solenoids* in generators used in hybrid and electric vehicles, as well as those in high-performance wind turbines, resulting in greatly increased efficiency. Permanent magnet synchronous motors are used in Honda, Toyota, Nissan, and Lexus hybrid vehicles and will most likely be the technology of choice for future electric vehicles.

While NdFeB and SmCo alloys were invented in the United States, production of these crucial materials has in the last decade been outsourced almost exclusively to China. That country's extensive mining operations currently

(continues)

Neodymium magnets are powerful enough to greatly improve the efficiency of alternative energy sources. ***(Splarka)***

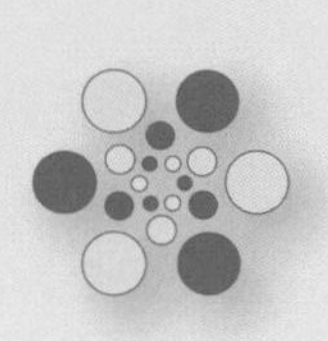

(continued)

provide nearly 100 percent of the world's rare earth elements. Demand worldwide is steadily increasing while Chinese exports of rare earth oxides are diminishing, owing to its focus on internal development of alternative energy sources. Estimates of the needs of the global energy industry indicate that annual neodymium production will need to at least double to keep pace with demand. Even the reopening of the Mountain Pass mine, which plans to produce 20,000 tons per year of rare earth metals (about 20 percent of current market demand) by 2012, will not come close to meeting this target.

Sm^{2+} and Yb^{2+} are easily oxidized in acidic solutions, as shown in the following reaction using Sm^{2+} as an example:

$$2\ Sm^{2+}\ (aq) + 2\ H_3O^+\ (aq) \rightarrow 2\ Sm^{3+}\ (aq) + 2\ H_2O\ (l) + H_2\ (g),$$

where "H_3O^+" is called the *hydronium ion,* which is a common way of depicting a hydrogen ion (H^+) in water. This reaction is especially favorable in the presence of dissolved oxygen, as showing in the following equation:

$$4\ Sm^{2+}\ (aq) + 4\ H_3O^+\ (aq) + O_2\ (g) \rightarrow 4\ Sm^{3+}\ (aq) + 6\ H_2O\ (l).$$

To say that a reaction is more favorable means that it proceeds at a more rapid rate.

Promethium needs to be treated separately from the other rare earths. Because all of the isotopes of promethium are radioactive, promethium's chemistry has been studied to a lesser degree than the chemistry of the other elements. To date, approximately 30 compounds of promethium have been made. All of its compounds have exhibited the

"+3" state. Examples of compounds that have been synthesized include promethium hydroxide [$Pm(OH)_3$], promethium chloride ($PmCl_3$), promethium nitrate [$Pm(NO_3)_3$], promethium oxide (Pm_2O_3), promethium carbonate [$Pm_2(CO_3)_3$], and promethium phosphate ($PmPO_4$). Because of promethium's radioactivity, all promethium compounds glow in the dark, emitting blue or yellowish green light.

TECHNOLOGY AND CURRENT USES OF THE LANTHANIDES

Currently, more than 85 percent of rare earth production goes into applications in metallurgy, the production of glass and ceramics, and catalysis. Rare earths are used in lasers, and the addition of rare earth elements gives mercury streetlights a bright white color. (When pure mercury is used, mercury streetlights are mostly purple in color.) The application of rare earths to crops has been demonstrated to increase productivity.

One of the most important applications of rare earths is in their combination in alloys called *mischmetals.* Mischmetal has a variety of uses. For example, it is used to remove oxygen from vacuum tubes and as a magnesium alloy to give magnesium greater strength. Natural combinations of the rare earths in mischmetal are used in cigarette lighters because they are ignited by friction and burn in air. More than 50 percent of rare earth production goes into the steel industry. In nonferrous metal applications, rare earths are alloyed with aluminum and silicon to make pistons and with zinc to make corrosion-resistant coatings.

The oxides of the rare earths have excellent abrasive properties and are used to polish mirrors and the glass surfaces of televisions. Rare earths can be used both to color glass and to decolor it. Compounds of praseodymium and neodymium are used to color porcelain products. In addition, rare earths are used to make glass for sunglasses that absorb light.

Chlorides of the rare earths are used as catalysts in the petroleum refining industry. Petroleum as it comes from the ground consists of a large mixture of compounds with a wide range of combustion properties. The refining process increases the fraction of desirable compounds.

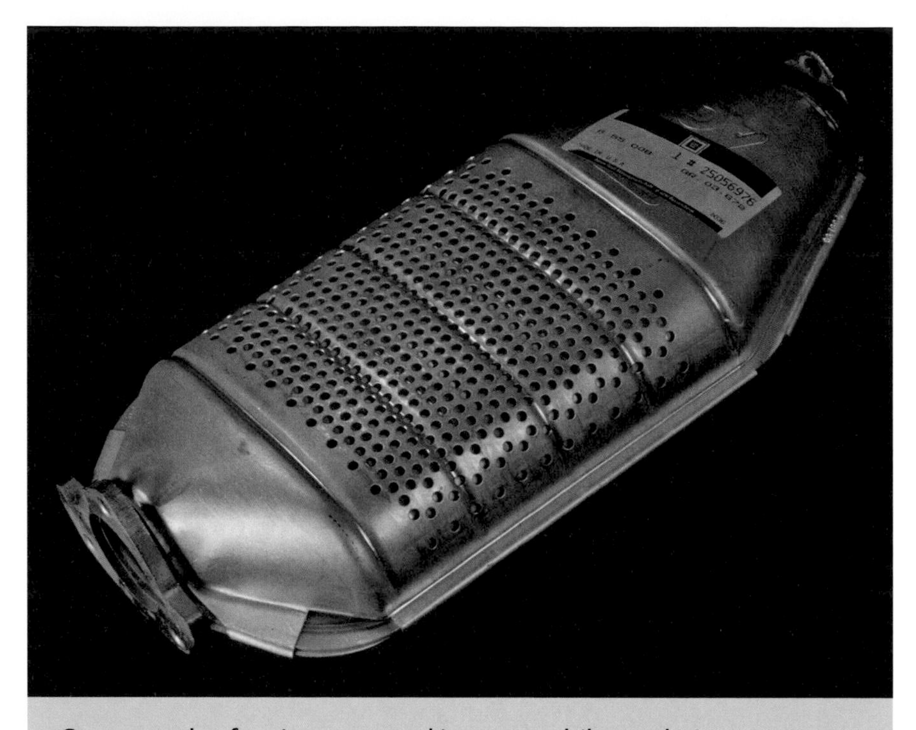

Compounds of cerium are used in automobile catalytic converters. *(Astrid & Hanns-Frieder Michler/Photo Researchers)*

Rare earth compounds have medical uses that date back to the middle 1800s. For example, rare earth compounds can be used to treat burns and as *antiseptic* agents. Cerium oxalate is used to treat motion sickness, vomiting, and nausea, and also helps to reduce the adverse effects of general anesthetics and chemotherapy.

Incandescent gas mantles contain cerium oxide. The oxide is also used to coat the walls of self-cleaning ovens to inhibit food particles from sticking to the walls. Compounds of cerium find many uses, including glass manufacturing, petroleum refining, automobile exhaust catalysts, and motion picture lighting.

Some alloys containing neodymium and samarium make excellent magnets. Europium and yttrium oxides impart the red colors to television pictures. Gadolinium is used in television and X-ray screens to increase fluorescence. Gadolinium is exceptional in that it is a *ferromagnetic* material. Injections of gadolinium salts into a patient enhance magnetic resonance images (MRIs). It also has the strongest *neutron capture* capability of any element and can be used in control rods in nuclear reactors. The metal is used in superconductors. Small amounts of gadolinium are added to metals such as iron and chromium to improve their workability. Erbium is used in optical fibers. It is also alloyed with vanadium to reduce its hardness. Erbium oxide is pink and is used to color glass and porcelain.

The most important use of rare earth metals in the near future appears to be in the so-called clean energy technologies, especially in their use as magnets in wind turbines and electric vehicle motors. Unfortunately, the mining of these materials has serious environmental effects that must be closely monitored, and recycling of spent products is crucial.

2

Actinium, Thorium, and Protactinium

Thorium (element 90), protactinium (91), uranium (92), and elements 93–103 are referred to as actinide elements because they follow actinium (element 89) in the periodic table. Just as the lanthanides (elements 58–71) are all considered to have chemical and physical properties that are similar to lanthanum (element 57), the actinides all have chemical and physical properties that are similar to actinium. In particular, the outermost electrons in elements 90–103 occupy "5f" orbitals (just as the outermost electrons in elements 58–71 occupy "4f" orbitals).

Of the three elements, thorium exists in greatest relative abundance. Thorium's principal ore is monazite, of which 10 percent consists of thorium with the remaining metal content consisting mostly of cerium, lanthanum, and neodymium. Actinium and protactinium tend to be found in uranium ores (see chapter 3).

In this chapter, the discovery, properties, and uses of actinium, thorium, and protactinium are treated separately from the other actinides. These three elements and uranium are the heaviest elements that occur naturally on Earth. Because of uranium's importance, that element is covered in its own chapter.

THE BASICS OF ACTINIUM

Symbol: Ac
Atomic number: 89
Atomic mass: 227.028
Electronic configuration: [Rn]$7s^2 6d^1$

T_{melt} = 1,922°F (1,050°C)
T_{boil} = ~5,788°F (3,198°C)

Abundance
In Earth's crust negligible

Actinium
2 8 18 32 18 9 2
Ac 89
1051°
3198°
+3
[227]

Isotope	Z	N	Relative Abundance	Half-Life
$^{227}_{89}Ac$	89	138	----	21.77 years

THE BASICS OF THORIUM

Symbol: Th
Atomic number: 90
Atomic mass: 232.038
Electronic configuration: [Rn]$7s^2 6d^2$

T_{melt} = 3,182°F (1750°C)
T_{boil} = ~8,650°F (4,788°C)

Abundance
In Earth's crust 6.0 ppm

Thorium
2 8 18 32 18 10 2
Th 90
1750°
4788°
+4
232.0381
1.09×10^{-10}%

Isotope	Z	N	Relative Abundance	Half-Life
$^{232}_{90}Th$	90	142	100%	14 billion years

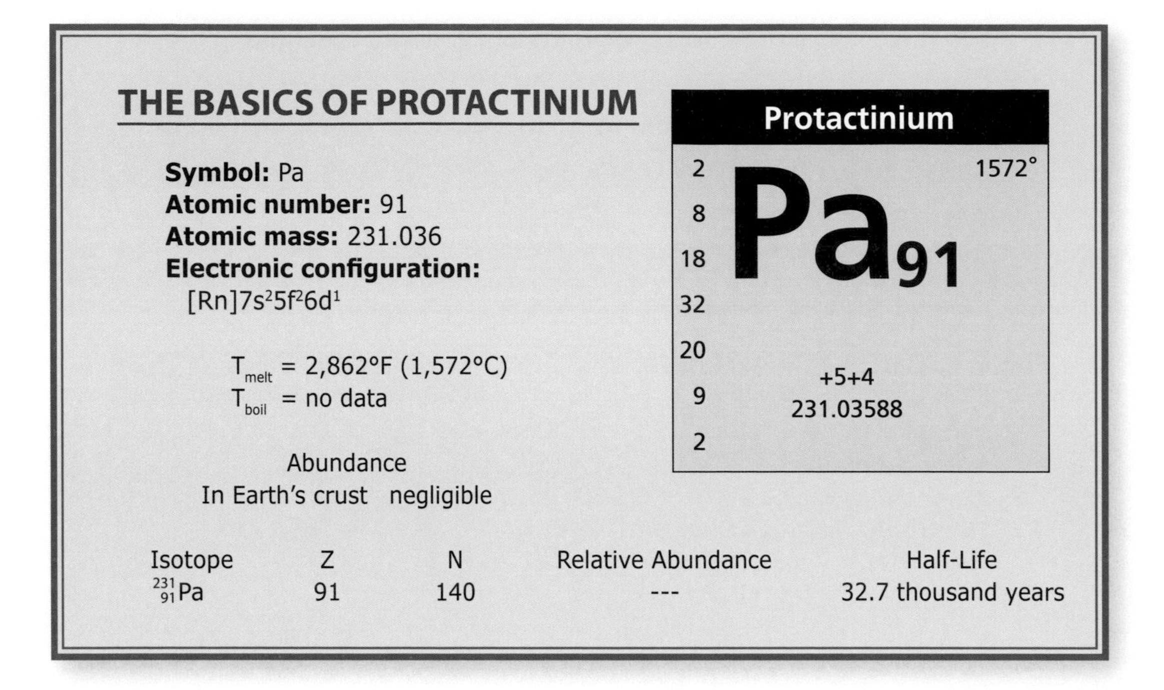

THE BASICS OF PROTACTINIUM

Symbol: Pa
Atomic number: 91
Atomic mass: 231.036
Electronic configuration:
$[Rn]7s^2 5f^2 6d^1$

T_{melt} = 2,862°F (1,572°C)
T_{boil} = no data

Abundance
In Earth's crust negligible

Isotope	Z	N	Relative Abundance	Half-Life
$^{231}_{91}Pa$	91	140	---	32.7 thousand years

THE ASTROPHYSICS OF THE ACTINIDES

All elements heavier than the lanthanides are manufactured in supernova explosions via the rapid successive capture by lighter nuclei of free-flying neutrons released in the blast, which is termed the r-process. In stellar spectra, however, almost none of the actinides or trans-actinides are observable because of their short (in the astrophysical sense) radiative decay *half-lives.* They have simply become other elements since the time they formed.

Especially in the spectra of the oldest known stars, only the nuclei of the heavier elements thorium 232 and uranium 238—with half-lives of 14 billion and 4.5 billion years, respectively—might still exist in high enough numbers for their *emission spectra to* be observable. In fact, the ratio of these isotopes in stars is known as a *cosmochronometer* because it can inform astrophysicists of the age of stars in which this ratio can be measured. From the time the star is born, these elements continue to decay with different half-lives. The isotopic ratio changes accordingly and can, thus, be used to determine the age of a star or even an entire galaxy (by taking the average value of a large sample of similar stars). By this method, some *galactic halo* stars have been calculated to be

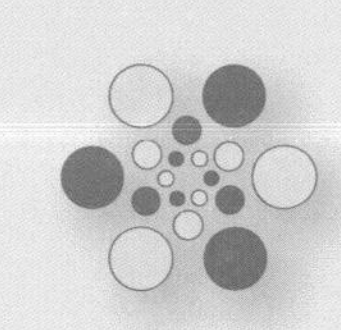

A MEANS OF DATING VERY OLD THINGS

Thorium 230 dating, also called uranium-thorium dating, is a method for determining the age of objects up to 500,000 years old. It is based on the ratio between the amount of thorium 230 and uranium 234 present in the object in question and cannot determine the age of every sort of material, only those that spent their lifetimes in sea water or other underground aquifers containing uranium. Many cave stalactites and stalagmites and marine organisms, such as ocean corals and fish fossils, are successfully dated using this method. Even the period of time volcanic particles have spent in the atmosphere can be determined by thorium-dating of raindrops.

Because uranium is highly soluble in water, while thorium is not, these materials absorbed significant uranium while in the water, but no thorium. Thorium 230 is, however, a radioactive

(continues)

Many cave stalactites and stalagmites can be successfully dated using thorium dating. *(Peter Essick/ Aurora/Getty Images)*

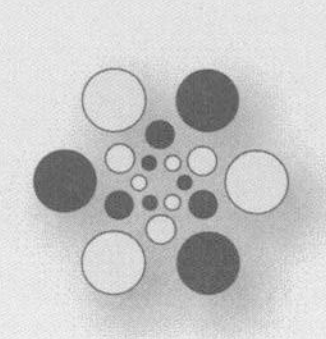

(continued)

daughter of uranium 234, as diagrammed in the figure. So over time (hundreds of thousands of years), ^{230}Th increases in the sample while ^{234}U decreases. A comparison of the amount of each isotope, therefore, yields the age of the sample.

The procedure only makes sense if the object of interest no longer was able to absorb uranium at a particular point. For volcanic particles, this happens at the time they are ejected from beneath the Earth's surface into the atmosphere. For coral or other marine organisms, it could be when the sea receded or when the dead sea life precipitated to the ocean floor and was covered by other sediments.

In such cases, thorium-dating is very accurate and has proven to be a good check on some objects for which carbon dating had been erroneous. Scientists typically use a process called *mass spectrometry* to separate the isotopes for analysis.

$^{238}_{92}U \xrightarrow{\alpha} {}^{234}_{90}Th$

$^{234}_{90}Th \xrightarrow{\beta} {}^{234}_{91}Pa$

$^{234}_{91}Pa \xrightarrow{\beta} {}^{234}_{92}U$

$^{234}_{92}U \xrightarrow{\alpha} {}^{230}_{90}Th$

$^{230}_{90}Th \xrightarrow{\alpha} {}^{226}_{88}Ra \xrightarrow{\alpha} {}^{222}_{86}Rn \xrightarrow{\alpha} {}^{218}_{84}Po \xrightarrow{\alpha} {}^{214}_{82}Pb$

$^{218}_{84}Po \xrightarrow{\beta} {}^{218}_{85}At$; $^{214}_{82}Pb \xrightarrow{\beta} {}^{214}_{83}Bi$

$^{218}_{85}At \xrightarrow{\alpha} {}^{214}_{83}Bi \xrightarrow{\alpha} {}^{210}_{81}Tl$

$^{218}_{85}At \xrightarrow{\beta} {}^{218}_{86}Rn$; $^{214}_{83}Bi \xrightarrow{\beta} {}^{214}_{84}Po$; $^{210}_{81}Tl \xrightarrow{\beta} {}^{210}_{82}Pb$

$^{218}_{86}Rn \xrightarrow{\alpha} {}^{214}_{84}Po \xrightarrow{\alpha} {}^{210}_{82}Pb$

$^{210}_{82}Pb \xrightarrow{\beta} {}^{210}_{83}Bi$

$^{210}_{83}Bi \xrightarrow{\alpha} {}^{206}_{81}Tl$

$^{210}_{83}Bi \xrightarrow{\beta} {}^{210}_{84}Po$; $^{206}_{81}Tl \xrightarrow{\beta} {}^{206}_{82}Pb$

$^{210}_{84}Po \xrightarrow{\alpha} {}^{206}_{82}Pb$

The decay of uranium 234 to thorium 230 is one of the earlier stages of the long decay chain from uranium 238 to lead 206.

between 11 and 15 billion years old. The *globular cluster* M15, which is within the Milky Way galaxy and appears in the constellation Pegasus, seems to have formed 14±4 billion years ago, while the method shows the Milky Way itself may be younger—around 12.5 billion years old.

As with *carbon dating*, the initial amount of each isotope needs to be known for the method to be accurate. Unfortunately, the field of *cosmochronology* must rely largely on theoretical models rather than astrophysical data for the initial values, a situation that is less than ideal. Further observations and a greater understanding of supernova events (and, thus, the r-process), will move the field forward.

THE DISCOVERY AND NAMING OF ACTINIUM, THORIUM, AND PROTACTINIUM

The phenomenon of radioactivity was discovered by the French physicist Henri Becquerel (1852–1908) in 1896. Because alpha and beta decay result in the formation of an element that differs from the parent element that underwent decay, scientists at first made the erroneous assumption that a rather large number of new elements heavier than radium (element 88) were being discovered among the radioactive decay products. For example, different decays would yield alpha particles possessing different kinetic energies, so it was thought that each value of kinetic energy was the result of a different element having undergone decay. The problem, however, was that thorium and uranium were well-known elements, and there did not appear to be enough space in the periodic table to crowd an array of new elements between or around them. It was not until the English chemist Frederick Soddy (1877–1956) developed the concept of isotopes in 1913 that scientists finally understood that samples of an element could consist of a variety of isotopes with each isotope possessing its own decay characteristics. Coupled with Henry Moseley's discovery of atomic number in the same year, it became quite clear that only three elements could exist that would lie between radium and uranium in the periodic table—actinium, thorium, and protactinium.

Thorium (element 90) was the first of these elements to have been discovered. In fact, its discovery occurred in 1829, several decades before the discoveries of the other two. Jöns Jacob Berzelius was given a black mineral that had been discovered on an island near Norway. The discoverer of the mineral, the Reverend Hans Esmark (1801–82) proposed to name the mineral *berzelite* in honor of Berzelius. Berzelius, however, chose to call it *thorite* and the new element that it contained *thorium,* named after Thor, the Scandinavian god of war. At the time that Berzelius discovered thorium, there was, of course, no lanthanide or actinide concept. For more than a century, chemists assumed that thorium belonged to the family of elements that included titanium and zirconium (hafnium was not discovered until the 1920s). It was not until the 1940s that the actinides were placed in their own row under the lanthanides. In addition, no one in Berzelius' day would have suspected thorium's radioactive properties since radioactivity was not discovered until almost 70 years later. Scientists now know that thorium has no stable isotopes. However, its longest-lived isotope has a half-life greater than the age of Earth, so that substantial primordial thorium still exists, usually in association with other rare earth minerals.

Actinium was discovered in 1899. *(Visuals Unlimited)*

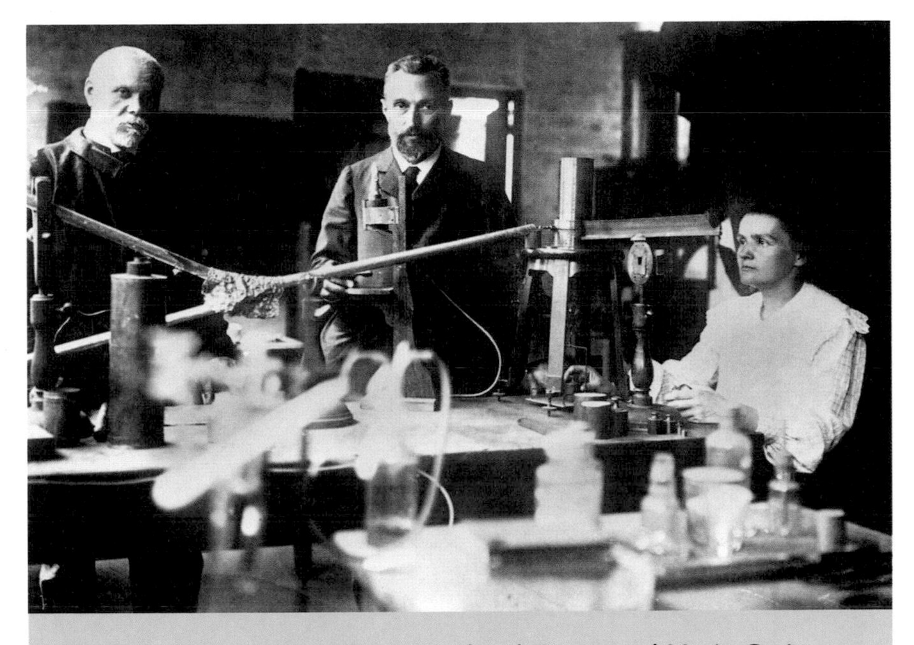

André Debierne, left, collaborated with Pierre and Marie Curie (1905). *(©Jacques Boyer/Roger-Viollet; l'Agence Roger-Viollet)*

Actinium (element 89) was not discovered until 1899. The Polish-French chemist Marie Skłodowska Curie (1867–1934) spent several years extracting radium from pitchblende, one of the principal uranium-containing ores. The French chemist André-Louis Debierne (1874–1949) worked closely with Curie, especially after the death of her husband, Pierre. Debierne continued the investigation of pitchblende and discovered a new radioactive element in it that he named *actinium* from the Greek word for "ray." Unaware of Debierne's discovery, the German chemist Friedrich Oskar Giesel (1852–1927) independently discovered actinium in 1902. Giesel wanted to call the new element *emanium* for the rays it emanated. Debierne, however, had priority of discovery.

The discovery of protactinium (element 91) was reported in 1917 simultaneously and independently by scientists in three separate locations: Otto Hahn (1879–1968) and Lise Meitner (1878–1968) in Berlin, Germany; Kazimierz Fajans (1887–1975) in Karlsruhe, Germany; and Frederick Soddy, John Arnold Cranston (1891–1972), and Alexander Fleck (1889–1968) in Glasgow, Scotland. Originally the new element was named *protoactinium,* meaning "first form of actinium" or "what comes before actinium," since actinium is the product formed when protactinium undergoes alpha decay. The first sample of the pure metal was prepared in 1934, and in 1949, the International Union for Pure and Applied Chemistry (IUPAC) shortened the name to protactinium. Before the actinides were placed in their own row of the periodic table, protactinium was believed to be a member of the vanadium family and was placed beneath tantalum.

THE CHEMISTRY OF THORIUM

Although all of thorium's isotopes are radioactive, thorium 232 has an exceptionally long half-life of 14 billion years (roughly the age of the universe). Since Earth is about 4.6 billion years old, that means that most of the primordial thorium that was present on Earth at its formation is still present. The most important sources of thorium are the minerals thorite and monazite. Thorium is often obtained as a by-product of uranium and rare earth mining.

Thorium compounds tend to occur only in the "+4" oxidation state. Examples are thorium chloride ($ThCl_4$), thorium fluoride (ThF_4), thorium dioxide (ThO_2), and thorium nitride (Th_3N_4).

The nuclear properties of thorium are probably more important than its chemical properties. Thorium is not sufficiently fissionable to be used as a nuclear fuel. However, in a nuclear reactor, it can be converted into uranium, which is fissionable. The process for producing uranium is shown in the following series of reactions:

$$^{232}_{90}\mathrm{Th} + {}^{1}_{0}\mathrm{n} \rightarrow {}^{233}_{90}\mathrm{Th} + \gamma \text{ (where } \gamma \text{ is a gamma ray);}$$

$$^{233}_{90}\mathrm{Th} \rightarrow {}^{233}_{91}\mathrm{Pa} + \beta;$$

$$^{233}_{91}\mathrm{Pa} \rightarrow {}^{233}_{92}\mathrm{U} + \beta.$$

U-233 is fissionable. Converting thorium into uranium for nuclear reactors extends the world's reserves of nuclear fuel resources.

Thorium 232 has a radioactive decay series. (See chapter 3 for uranium's radioactive decay series.) Ten radioactive decays—a combina-

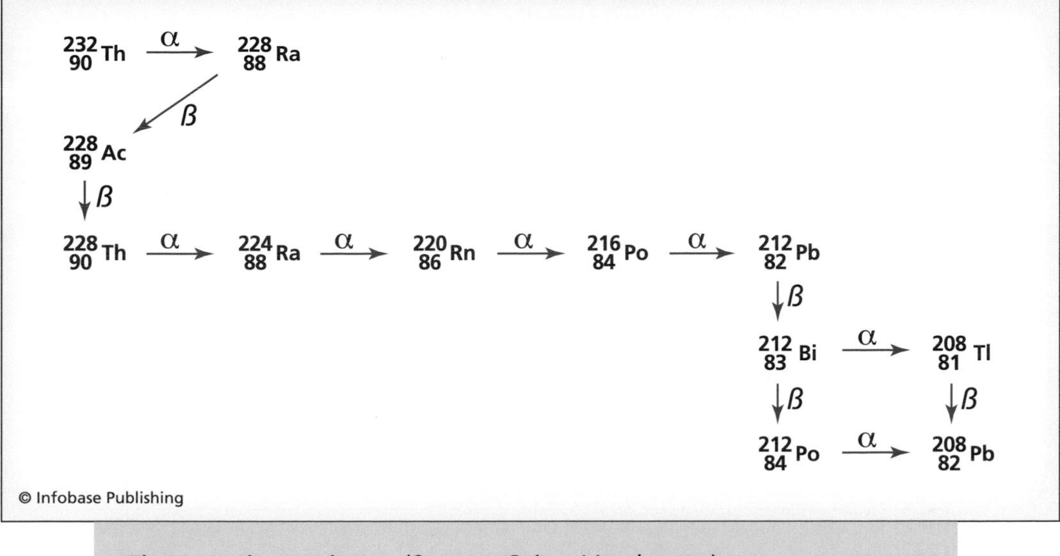

Thorium decay chain *(Source: Brian Nordstrom)*

tion of alpha (α) and beta (β) decays—take place, eventually resulting in the formation of lead 208, which is stable. Each alpha decay yields a daughter that is two spaces to the left of the parent in the periodic table; the mass number decreases by four units. Each beta decay yields a daughter that is one space to the right of the parent; the mass number does not change. When bismuth 212 is reached, both alpha and beta decay can occur. With the exception of thorium 232 itself, the half-lives of the decays are all very short. The isotope with the shortest half-life, 30 microseconds, is polonium 212. The longest half-life, 6.7 years, is that of radium 228. The decay series is shown above.

THORIUM FISSION REACTORS

Thorium-fueled *fission reactors* could, in the long term, provide a large fraction of the energy demand of the planet more efficiently and with fewer hazardous side effects than current-day uranium enrichment reactors. Uranium reactors (see chapter 3) rely on slow neutrons to trigger the fission event. While natural thorium (most commonly found as the thorium 232 isotope) itself is not fissile, the absorption of a slow neutron by Th-232 begins a series of *beta decay* reactions that result in uranium 233, which is a good nuclear fuel. The reactions involved are as follows:

$$^{232}_{90}\text{Th} + ^{1}_{0}\text{n} \rightarrow ^{233}_{90}\text{Th} + \beta^{-}$$

$$^{233}_{90}\text{Th} \rightarrow ^{233}_{91}\text{Pa} + \beta^{-}$$

$$^{233}_{91}\text{Pa} \rightarrow ^{233}_{92}\text{U} + \beta^{-}$$

This series proceeds with an efficiency of about 90 percent: Ten percent of the time protactinium 233 (^{233}Pa) absorbs a neutron to form uranium 234. The fission of uranium 233 can then be triggered by slow neutrons, such as are intentionally produced in reactors to instigate the fission of uranium 235. The science is well understood for this type of reactor—Oak Ridge National Laboratory successfully tested one such device, called the Molten Salt Experiment, for a few years in the late 1960s. That was the time of the *cold war,* however, when plutonium stockpiles for weapons were a desirable by-product of reactors that also produced electrical power, so uranium 235 remained the fuel of choice. (The uranium 233 decay chain does not produce plutonium.)

Uranium is now in short supply and commanding a much higher price than in previous decades; the price of a pound of uranium increased 20-fold from 2001 to 2006. Additionally, fears of plutonium getting into the wrong hands have made the thorium reactor option more attractive. Thorium is widely available in many countries in deposits that require less environmentally destructive mining techniques than does uranium ore. Radioactive waste by-products from thorium reactors would have short half-lives (on the order of decades), so safe storage would not be onerous, and bomb production from the waste products would be much more difficult and expensive than such production from plutonium.

The switch to thorium is not simple, however. Companies will need an incentive to invest in mining. The push will most likely not come from power companies but from the public sector as people learn that this option is available. In April 2008, a *heavy water* test reactor that incorporates *passive* safety features was initiated in India, which plans to rely on such reactors for a large proportion of its power in the near future. In March 2009, the U.S. Congress advised the investigation of thorium-based fuels for nuclear submarines. Legislation introduced in 2008 by Senators Orrin Hatch and Harry Reid, dubbed the Energy

Independence and Security Act, which recommends exploration of the use of thorium reactors as a method of assuring energy security has been tabled. Meanwhile, research in this area continues in several countries that can boast known thorium resources, such as India, Norway, Russia, and Canada.

TECHNOLOGY AND CURRENT USES OF ACTINIUM, THORIUM, AND PROTACTINIUM

Actinium is important to researchers as a source of neutrons and as a tracer to study deep sea water mass movement. Its only commercial use is in satellites, which can take advantage of the energy stored in its radioactive decay chain for thermoelectric power generation. Actinium has no widespread consumer application at this time.

Thorium is a different story. This element in its most available form may within decades turn out to be the world's most important power source. Thorium-fueled fission reactors could, in the long term, provide a large fraction of the energy demand of the planet more efficiently and with fewer hazardous side effects than current-day uranium enrichment reactors. Research in this area is ongoing in several countries that can boast known thorium resources, especially India, but also Norway, Russia, and Canada. (The U.S. Congress has recently taken some small steps in this direction, such as instructing the navy to look into powering its submarines with thorium.) Currently, thorium oxide has several consumer applications, including its use in high-refractive-index glass for optical faithfulness in lenses for scientific instruments, in lamps for brightness, and as a catalyst in petroleum and sulfuric acid production.

Like actinium, protactinium is currently most important in scientific research, where it has been used as a tracer in oceans to help understand the movement of North Atlantic streams during glacial melt. Given the current global warming issues, this element may serve as a marker for further glacial melt.

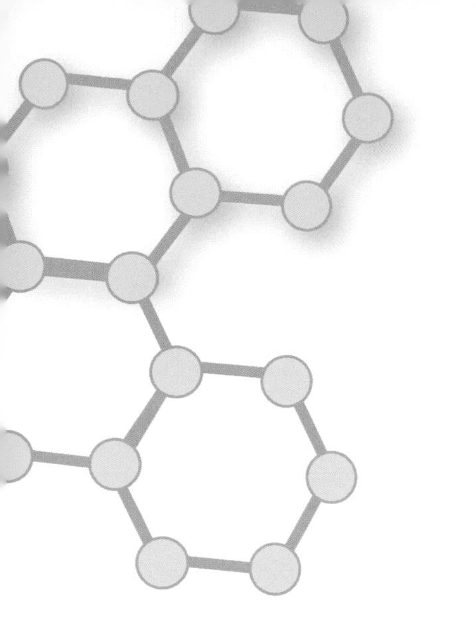

3

Uranium

Uranium is element 92 and is the heaviest naturally occurring element. In fact, when Dmitri Mendeleev published his periodic table of the elements in 1869, there were no artificial elements, and uranium was believed to be the end of the table. It was only in the 1930s that scientists began to even search for possible heavier elements, a search that resulted in the discovery of the first so-called *transuranium* element, neptunium, in 1940. Today, the discoveries of a full 25 transuranium elements have been reported, and the search for heavy elements continues to be an active area of scientific activity.

The word *uranium* generates a mixture of emotions among the general public. Most people know that uranium is radioactive and that prolonged exposure to it (as occurs among workers in uranium mines) can produce severe adverse health effects. Most people also know that uranium is the fuel used in nuclear power plants and that uranium can be used to make atom bombs. Varying emotions arise because of perceptions about nuclear power. People who support the benefits of nuclear power tend to view uranium as contributing to the common good. However, many people fear the risks of nuclear power because of their dread of uranium's radioactivity and oftentimes because of misconceptions about the overall risks associated with nuclear power plants. In addition, 50 years of cold war tensions that existed between the United States and the former Soviet Union—creating the real possibility of a devastating nuclear confrontation between the two powers—also contributed to people's fear of anything nuclear. (That fear persists today with the specter of terrorists and hostile foreign governments.)

The purpose of this chapter is to present a balanced treatment of uranium and its history, chemistry, and role in power plants and weapons technology.

THE BASICS OF URANIUM

Symbol: U
Atomic number: 92
Atomic Mass: 238.029
Electronic Configuration:
$[Rn]7s^2 5f^3 6d^1$

T_{melt} = 2,070°F (1,132°C)
T_{boil} = 6,904°F (3,818°C)

Abundance
In Earth's crust 1.8 ppm

Uranium

2, 8, 18, 32, 21, 9, 2

U_{92}

1135°
4131°

+3+4+5+6
238.0289
2.94×10^{-11}%

Isotope	Z	N	Relative Abundance	Half-Life
$^{234}_{92}U$	92	142	0.0054%	248 thousand years
$^{235}_{92}U$	92	143	0.71%	713 million years
$^{236}_{92}U$	92	144	---	23.4 million years
$^{238}_{92}U$	92	146	99.28%	4.51 billion years

THE GEOLOGY AND MINING OF URANIUM

The majority of uranium mined on Earth at this time is found in *unconformity, vein,* and *sandstone* deposits. The largest percentage comes from unconformities, which consist of *sedimentary* depositions between hard rock layers with different ages. To get at the uranium in these deposits generally involves open-pit mining. Heavy machinery is required to dig, remove, and break rock to get to the important layer or to carry ore that may have even a small uranium content to heaps or vats where it can be further refined into the usable uranium oxide end product (U_3O_8) called *yellowcake.*

The process of soil and bedrock removal can expose previously buried sulfur-containing minerals to air and weather. Rainfall or immersion of these minerals in streams produces sulfuric acid, which can dissolve heavy metals that are normally bound in soils. This *acid mine drainage* also has the effect of changing the pH of the water, which can be extremely detrimental to aquatic life. In addition, the dust raised during the moving of uranium-bearing rock contains radioactive *particulates,* which must not be allowed to enter the surrounding atmosphere. Typically, a large quantity of water is needed to spray down the dust as it

Uranium ore can resemble commonplace gray rock. *(Andrew Silver/USGS)*

An aerial view of an open pit uranium mine on the Spokane Indian Reservation near Wellpinit, Washington. Uranium ore was mined until 1981, when the mine closed. It is still in the process of reclamation. *(AP Photo/HO)*

rises. Clearly, windy conditions complicate the process. Unconformity deposits are especially rich in Australia, but Namibia, Niger, and Canada also mine uranium in this manner.

Vein deposits are mined in underground tunnels, where radon gas (a *daughter isotope* of uranium) can put miners at risk. Uranium particulates are also a problem in these mines—which provide about 10 percent of the world's uranium. (See "Uranium and Human Health" section later in this chapter.) Historically important veins were located in the Czech Republic and Democratic Republic of the Congo (formerly Zaire); the latter supplied uranium to the United States for the Manhattan Project, as did a sandstone deposit in the Northwest Territories of Canada. Currently, the highest-producing underground mines are in Canada, Australia, and Russia.

THE TOP 10 URANIUM-PRODUCING SITES*

MINE	COUNTRY	MAIN OWNER	TYPE	PRODUCTION (TONNES U)	PERCENT OF WORLD
McArthur River	Canada	Cameco	Underground	7,339	15
Ranger	Australia	ERA (Rio Tinto 68%)	Open pit	4,444	9
Rossing	Namibia	Rio Tinto (69%)	Open pit	3,520	7
Kraznoka-mensk	Russia	ARMZ	Underground	3,004	6
Olympic Dam	Australia	BHP Billiton	By-product/ Underground	2,955	6
Tortkuduk	Kazakhstan	Areva	In situ leaching	2,272	4
Arlit	Niger	Areva/ Onarem	Open pit	1,808	4
Rabbit Lake	Canada	Cameco	Underground	1,447	3
Akouta	Niger	Areva/ Onarem	Underground	1,435	3
Budenov-skoye 2	Kazakhstan	Kazatomo-prom	In situ leaching	1,415	3
Top 10 total				**29,638**	**59%**

*2009 data
Source: www.world-nuclear.org/info/inf23.html

Sandstone deposits, which supply about 25 percent of the global demand, are the only available sources of uranium in the United States. The recovery of uranium from sandstone is less detrimental to the natural environment than either open-pit or underground vein mining. *In situ leaching* (see figure on page 53) is the process of introducing an acid or alkaline solution (depending on the calcium concentration in the sandstone) in which the uranium dissolves, and the solution is then pumped out on the far side of the deposit. This process is much less expensive than open-pit or vein mining because rock does not have to

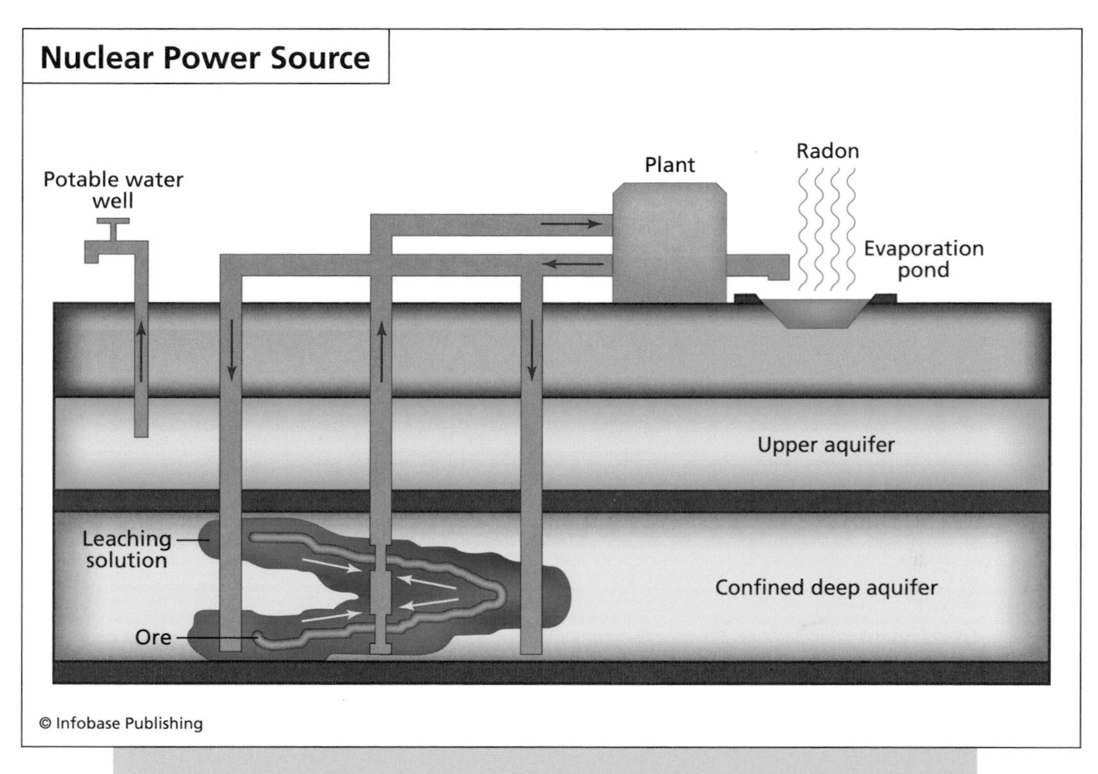

In situ leaching is the process of introducing a solution in which the uranium dissolves. The solution is then pumped out on the far side of the deposit.

be moved. It can only be used, however, when the surrounding material is porous enough for the liquid to move through the rock. The main hazard to this type of uranium extraction is the difficulty of determining the underlying geology, and radioactive fluid can leach into groundwater. Such concerns prompted U.S. Secretary of the Interior Ken Salazar in July 2009 to announce a two-year moratorium on new uranium claims in and around Grand Canyon National Park. An official environmental review is under way, which could result in a ban on new claims in that region for decades to come.

DISCOVERY AND NAMING OF URANIUM

The use of uranium oxide to impart a green color to glass dates back to the first century C.E. The discovery that uranium is an element, however, did not occur until the latter half of the 18th century.

Glowing uranium glass. Uranium oxide imparts a green color to glass, here shown fluorescing under ultraviolet light. *(Z. Vesoulis)*

One of the greatest German chemists of late 18th and early 19th centuries was Martin Heinrich Klaproth (1743–1817). At the time, pitchblende was a well-known ore that chemists believed to be a combination of zinc and iron. In 1789, when Klaproth analyzed pitchblende, he concluded that there was an unknown metallic element in it. Klaproth dissolved a sample of pitchblende in nitric acid. When he added potassium hydroxide, a yellow solid precipitated. When Klaproth heated the yellow solid with charcoal, he obtained a black powder that he believed to be a pure new element. In 1781, the English astronomer William Herschel (1738–1822) had discovered the planet Uranus, a discovery of great importance in astronomy because Uranus was the first planet to have been discovered since antiquity. In honor of Uranus, Klaproth named his new element *uranium*. It was not until 1841 that it was demonstrated that Klaproth had only isolated an oxide of uranium, not the pure metal itself. (The yellow solid most likely was sodium diuranate, $Na_2U_2O_7 \cdot 6H_2O$, a common form of uranium that is often referred to in the uranium mining industry as yellowcake.)

Despite his not having obtained a pure sample, Klaproth was still credited as the discoverer of uranium.

Klaproth is remembered for several important contributions to chemistry. He was one of the founders of the field of analytical chemistry. Additionally, he played a role in the discoveries of several elements besides uranium—zirconium, strontium, titanium, chromium, tellurium, and cerium (see chapter 1). Klaproth determined the contents of a number of minerals and contributed to early studies of the rare earths. Of course, the properties of uranium that are best known to the general public are its radioactivity and ability to undergo fission, but those properties were not discovered until many years after Klaproth's death. One can only speculate how much radiation exposure scientists like Klaproth experienced in the years before scientists discovered radioactivity and realized its health hazards. Since Klaproth lived to be 73 years old, it may be that he did not suffer ill effects from his handling of uranium, though the cause of his death is unknown.

THE CHEMISTRY OF URANIUM

There are a number of minerals that contain uranium. The most significant example is uraninite, of which pitchblende is one variety. The only important metal in uraninite is uranium. Other minerals typically contain several other metals in addition to uranium. Uraninite is the mineral commonly mined in the western United States. Other uranium-bearing minerals include euxenite, mined in Idaho; carnotite and coffinite, mined in Colorado; autunite, mined in France; uranophane, mined in Africa; and brannerite, mined in Canada. At least small amounts of uranium, however, are found in the majority of rocks that constitute Earth's crust. Granite, for example, often has uranium in it. Building materials may contain traces of uranium. Therefore, people are sometimes concerned about exposure to radiation resulting from housing materials and from the soils and rocks on which their houses have been built. Another often-voiced concern pertains to radon gas. Since radon is a product of the uranium decay series (see page 57) and is the only gaseous element in the series, there is the potential for inhaling radon, especially in basements and other enclosed areas. Radon is an alpha emitter. Alpha particles outside the body are dangerous, but inhaling radon emits alpha particles in the soft tissues of the lungs where there is the potential for damage. Studies of the dangers of radon gas in homes, however, have produced ambiguous conclusions about its hazards to human health.

The *ordinary* chemistry of uranium (that is, chemistry that does not involve nuclear reactions) is dominated by compounds or ions in which uranium is in the "+4" or "+6" oxidation states. It is very difficult to isolate pure uranium metal, as demonstrated by the inability of chemists to do so during Martin Klaproth's lifetime. The usual procedure is to use magnesium metal to reduce uranium tetrafluoride to uranium metal at high temperature as shown by the following reaction:

$$UF_4\ (l) + 2\ Mg\ (l) \rightarrow U\ (l) + 2\ MgF_2\ (l).$$

There is, however, little demand for metallic uranium. Applications of uranium usually involve its compounds.

Besides uranium tetrafluoride, uranium compounds in the "+4" oxidation state include uranium tetrachloride (UCl_4) and uranium dioxide (UO_2). Compounds in the "+6" state include uranium hexachloride (UCl_6), uranyl chloride (UO_2Cl_2), uranyl sulfide (UO_2S), uranium trioxide (UO_3), and uranium *hexafluoride* (UF_6). Uranium hexafluoride is an especially important compound. It can be heated easily to the gaseous state where it is used in the *gaseous diffusion* process to achieve separation of uranium isotopes. Gaseous diffusion makes use of the fact that molecules in the gaseous state travel at average speeds that are inverse functions of their masses. In other words, light molecules tend, on average, to travel at higher average speeds than heavier molecules do. Uranium's isotopes differ only slightly in mass, but the difference in mass is reflected in the masses of uranium's different isotopes in UF_6 molecules. Gaseous diffusion separation is a tedious process with a large number of steps, but the difference in molecular speeds can be used to separate the different isotopes.

The *nuclear* chemistry of uranium (that is, processes that result in *transmutations* of uranium into other elements) is dominated by two processes: radioactivity and nuclear fission. Radioactivity and uranium's *radioactive decay series* will be discussed first.

All of the elements heavier than bismuth exist only as radioactive isotopes. They all undergo a series of radioactive decays until they end up as isotopes of lead (element 82) or bismuth (element 83). Uranium has three principal naturally occurring isotopes: U-234, with a half-life of 248 thousand years; U-235, with a half-life of 713 million years; and U-238, with a half-life of 4.51 billion years. Almost all of the uranium (99.3 percent) is U-238.

U-238 and its daughters undergo 14 decays—a combination of alpha (α) and beta (β) decays—to end up as Pb-206, a stable isotope, as

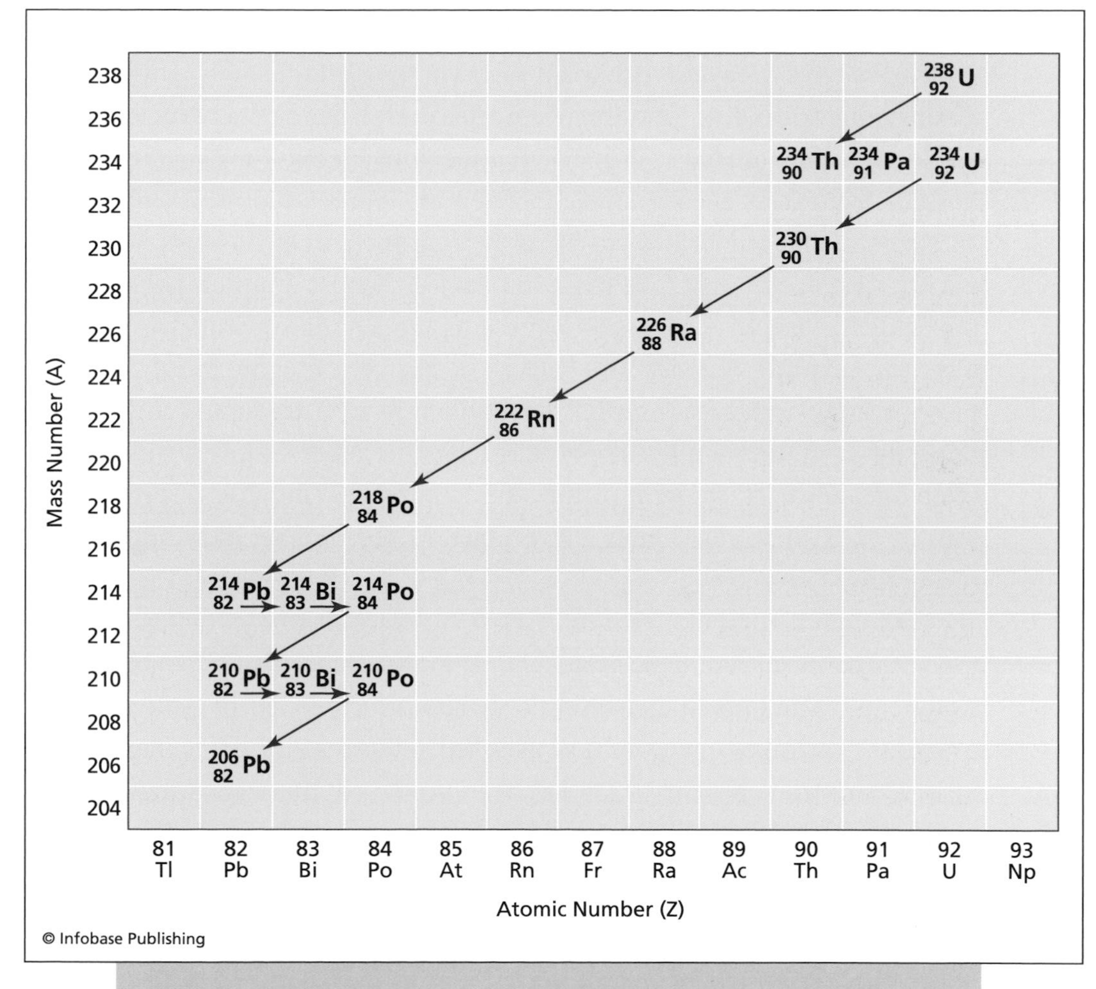

Uranium decay chain

shown in the diagram on page 57. Once polonium 218 is reached, there are isotopes that can decay by both modes, but everything eventually ends as Pb-206. The half-lives of decay can be as short as milliseconds, or—as in the case of U-238—as long as billions of years. Other isotopes of uranium have similar decay series. With the exception of thorium, the isotopes of all of the elements between bismuth and uranium have half-lives much shorter than the age of Earth. Therefore, no primordial quantities of those elements still exist. What small amounts are found in Earth's crust or mantle are there because they have been replenished by the uranium and thorium decay series. As they form, however, most of them decay away again quickly.

Radioactive decay can occur anywhere in the periodic table and is an unsymmetrical disintegration of a nucleus. Alpha and beta particles are very small compared to the nucleus undergoing decay. The daughter nucleus that forms is comparable in size to its parent and represents an element only one or two spaces away from the parent. In contrast, nuclear fission can only take place in heavy elements, uranium being the lightest example. Also, fission represents a much more symmetrical disintegration of a nucleus. The nucleus breaks very roughly into two pieces of comparable mass, so that the products formed are significantly removed from uranium in the periodic table. Because of conservation of charge, however, the sum of the atomic numbers (number of protons) of the two products must always equal 92, the number of protons in a uranium nucleus.

Although natural uranium consists of 99.3 percent U-238, that isotope of uranium is not fissionable. The naturally occurring fissionable isotope is U-235, which constitutes only about 0.7 percent of natural uranium. Natural uranium must be *enriched* to increase the percentage of U-235. Once sufficient U-235 has been produced, there are several pairs of products that can be formed from splitting its nuclei. The fission process begins with the absorption of a neutron by a uranium nucleus. Besides the two new elements that are produced, two to three neutrons are also released. Since those neutrons can be absorbed by surrounding uranium nuclei, more fission occurs. A *chain reaction* occurs as the rate of fission increases exponentially.

Examples of fission pathways include the following reactions:

$$^{235}_{92}U + ^{1}_{0}n \rightarrow ^{236}_{92}U \rightarrow ^{89}_{36}Kr + ^{144}_{56}Ba + 3\ ^{1}_{0}n;$$

$$^{235}_{92}U + ^{1}_{0}n \rightarrow ^{236}_{92}U \rightarrow ^{92}_{36}Kr + ^{142}_{56}Ba + 2\ ^{1}_{0}n;$$

$$^{235}_{92}U + ^{1}_{0}n \rightarrow ^{236}_{92}U \rightarrow ^{92}_{38}Sr + ^{141}_{54}Xe + 3\ ^{1}_{0}n.$$

One of the problems with nuclear reactors is that the fission fragments that are produced are all radioactive materials. Their half-lives are sufficiently long that they may be stored for thousands of years.

Although U-238 is not fissionable, it does absorb neutrons. The product, U-239, is a beta emitter, producing Np-239. Neptunium 239 is also a beta emitter, producing Pu-239. Plutonium 239 is fissionable. Because only 11 pounds (5 kilograms) of plutonium are required to fashion a nuclear warhead—versus 33 pounds (15 kilograms) of uranium—plutonium is the material of choice for nuclear weapons.

URANIUM FISSION REACTORS: NATURAL AND HUMAN-MADE

Nuclear fission reactors provide 14 percent of the world's electricity, 20 percent of electricity in the United States, and 100 percent of the plutonium used in nuclear weapons. Unlike weapons, however, reactors require a controlled scenario; heat energy must be produced without an explosive event. The heat is used to boil water to make steam to turn turbines that provide electricity to more than 100 million people worldwide.

The starting point for a functional reactor is achieving the correct abundance of uranium 235, which in natural occurrence is only 0.720 percent (except in fossil reactors, discussed below). The most abundant isotope of uranium is ^{238}U, which does not have the necessary properties to sustain the nuclear chain reaction under the circumstances most reactors operate. The desirable operating conditions include (1) the availability of slow neutrons to induce fission; (2) the ability of normal water to moderate the reaction and cool the reactor; and (3) maintenance of a constant neutron population at the core. To achieve these criteria, a minimal abundance of 3 percent uranium 235 is needed and can be attained by the process of *uranium enrichment,*

Nuclear power plants provide 14 percent of the world's electricity. *(Martin D. Vonka/Shutterstock)*

for which mass separation is the key: Because uranium 235 is lighter than uranium 238, it diffuses faster. In the gaseous diffusion process, *pressure differentials* are employed to draw the ^{235}U into a separate compartment. What is left over is depleted in U-235 as compared with natural uranium, so it is termed *depleted uranium,* which is discussed later in this chapter.

A newer, more energy-efficient method for separating isotopes relies on centrifugal action to separate the lighter U-235 from the heavier U-238, which propagates toward the outer rim of a rotating cylinder. Heat is sometimes then used to lift the desired U-235 up and away from the depleted gas. The centrifuge process currently accounts for more than half of the U-235 enrichment worldwide.

Once the correct fraction of U-235 is obtained, fuel elements are assembled and placed in a chamber of water or other *moderator.* Uranium 235 and 238 isotopes both spontaneously fission, exploding into smaller fragments—nuclei of lower atomic mass—but also emitting fast neutrons, heat, and gamma rays. The main advantage of U-235 over

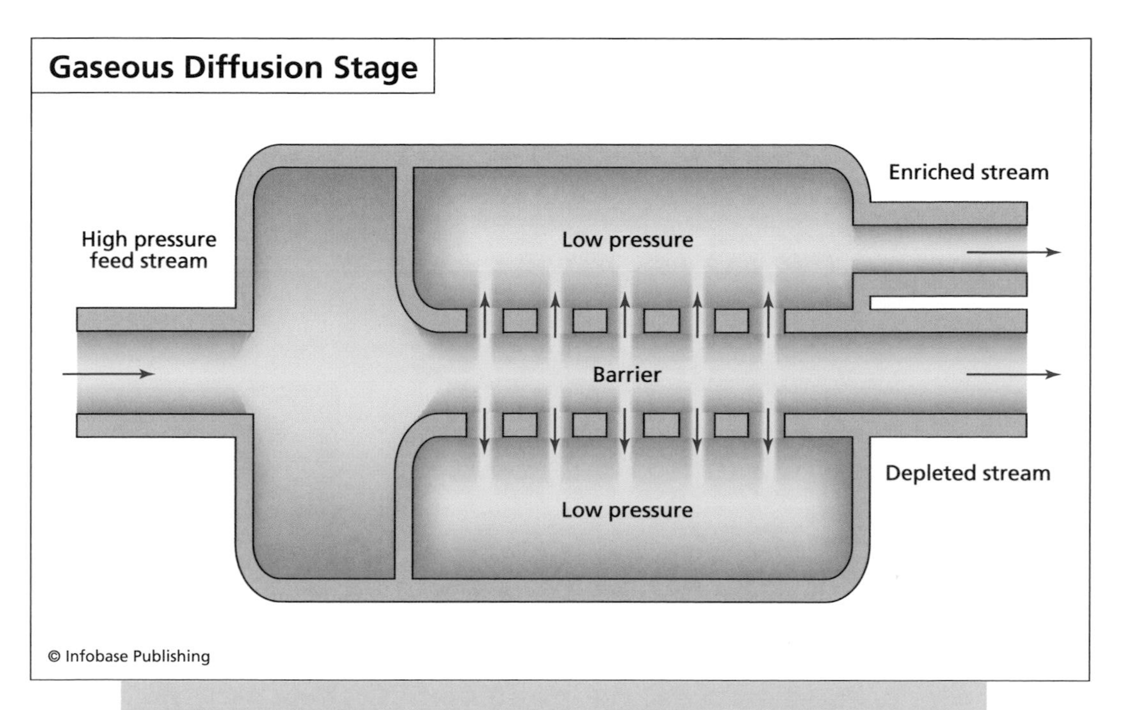

The gaseous diffusion process is one method for uranium enrichment. Uranium hexafluoride is diffused through a series of porous membranes. Velocity separation is the key: Because uranium 235 is lighter than uranium 238, it diffuses faster through the membranes. Pressure differentials are employed to draw the ^{235}U into a separate compartment.

U-238 is that fission can be induced in U-235 by collisions with neutrons, but not very efficiently with the naturally produced fast neutrons, which bounce around a lot and easily escape. These can be slowed down, however, by passing through water or molten salt—materials that are termed *moderators* and are crucial for the efficient operation of nuclear power sources.

A problem could result, however, if too many neutrons were available at any given time. This could lead to an exceedingly high number of fissions in the core and a runaway reaction or *supercriticality. Control rods* that can be inserted or removed solve this problem. They are typically made of materials such as silver, indium, cadmium, or boron that

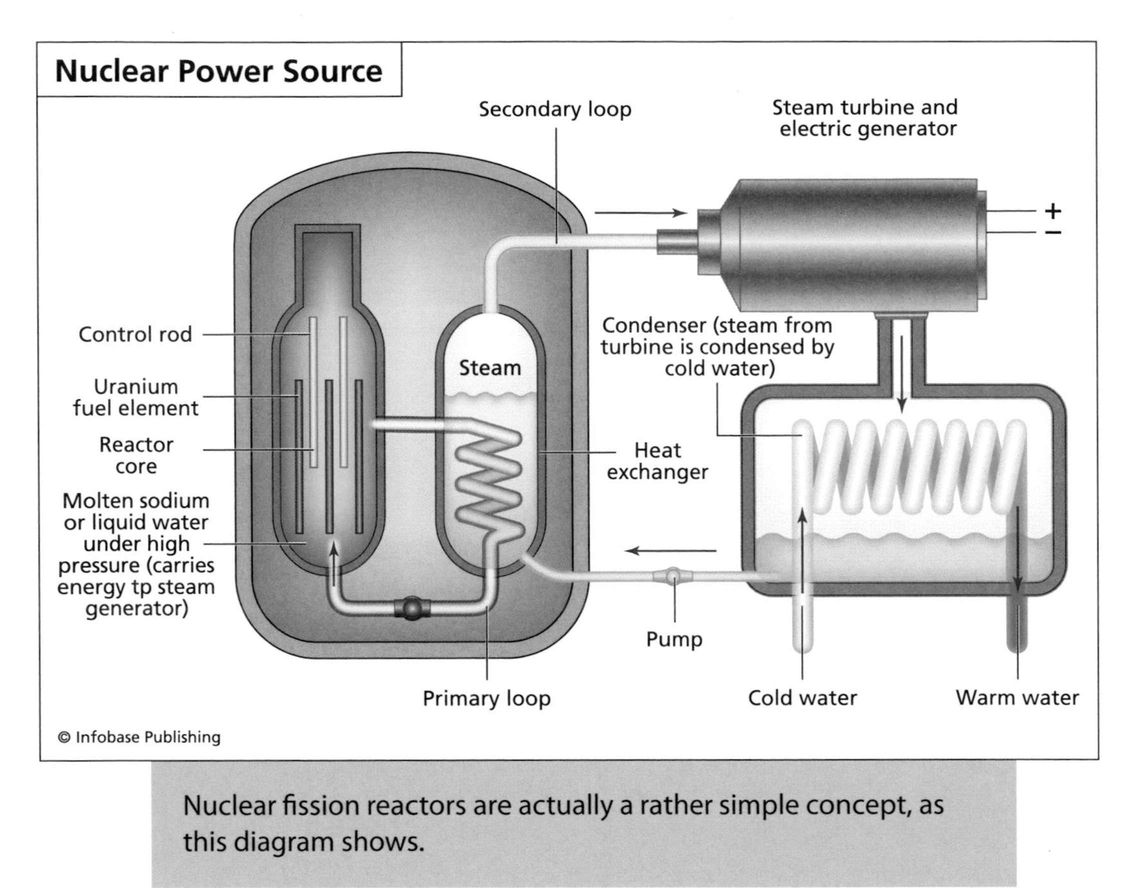

Nuclear fission reactors are actually a rather simple concept, as this diagram shows.

easily capture neutrons, thereby making them unavailable for fission reactions. The diagram above shows the main components of a nuclear reactor.

Surprisingly, in the distant past, Earth itself provided all the necessary components of a nuclear power reactor that sustained itself for tens of thousands of years. In what is now the African country of Gabon, a sandstone deposit with veins of sedimentary rock apparently achieved *criticality* about 1.7 billion years ago in at least 15 distinct sites in three different ore deposits. This was possible because at that time the U-235 abundance was higher, about the 3 percent necessary to trigger a reaction. And the conditions were right: Water flowing through the rock served as a moderator to allow the reactor to function at criticality in a periodic manner with astonishing regularity—30 minutes on and two and a half hours off. The reactors (now called the Oklo Fossil Reactors) turned themselves off when the water turned to steam, which is not capable of slowing down the fast neutrons produced by the spontaneous fission of uranium. When the temperature cooled and water continued to flow through the rock, the reactions could start again. A significant amount of plutonium 239 (^{239}Pu) would have been produced via the following reaction:

$$n + {}^{238}_{92}U \rightarrow {}^{239}_{92}U + \gamma \xrightarrow{23.5m} {}^{239}_{93}Np + e^{-} \xrightarrow{2.35d} {}^{239}_{94}Pu + e^{-}.$$

The half-lives of U-239 and Pu-239 are given respectively above the arrows where the radioactive decays occur. (*Antineutrinos* are also produced in the decays.) Plutonium 240 would have also formed when neutrons were absorbed by Pu-239. This means that plutonium, not uranium is the heaviest naturally produced element on Earth, though the abundance of ^{235}U in ore has been too low to have triggered such events for at least 1.5 billion years.

The United States has more nuclear reactors in operation (104) than any other country, with France and Japan running a close second at 59 and 54, respectively (see the figure on page 64). The need for reduction of carbon emissions combined with a greater measure of energy independence, however, has prompted plans for many more to be built in the near future. From 2007 to January 2010, the U.S. Nuclear Regulatory Commission received 21 new construction applications, for a total of 31 projected units (see map on page 65).

The nuclear industry has a couple of important problems to solve before it can really advance as a reliable and widespread source of electricity. (1) The public must be convinced that there is no risk of a runaway reaction that could cause a nuclear accident. (2) The industry must find acceptable sites to bury the radioactive waste or find some other means of alleviating its toxic potential. Possible solutions may lie in next-generation reactors that incorporate inherently safer features. Under ongoing investigation are the *Very-High Temperature Reactor* and the *Sodium-Cooled Fast Reactor*. Both will be much more efficient than current models, but neither could be available beyond the prototype before 2025. Thorium reactors may well turn out to be the best choice for future development (see chapter 2).

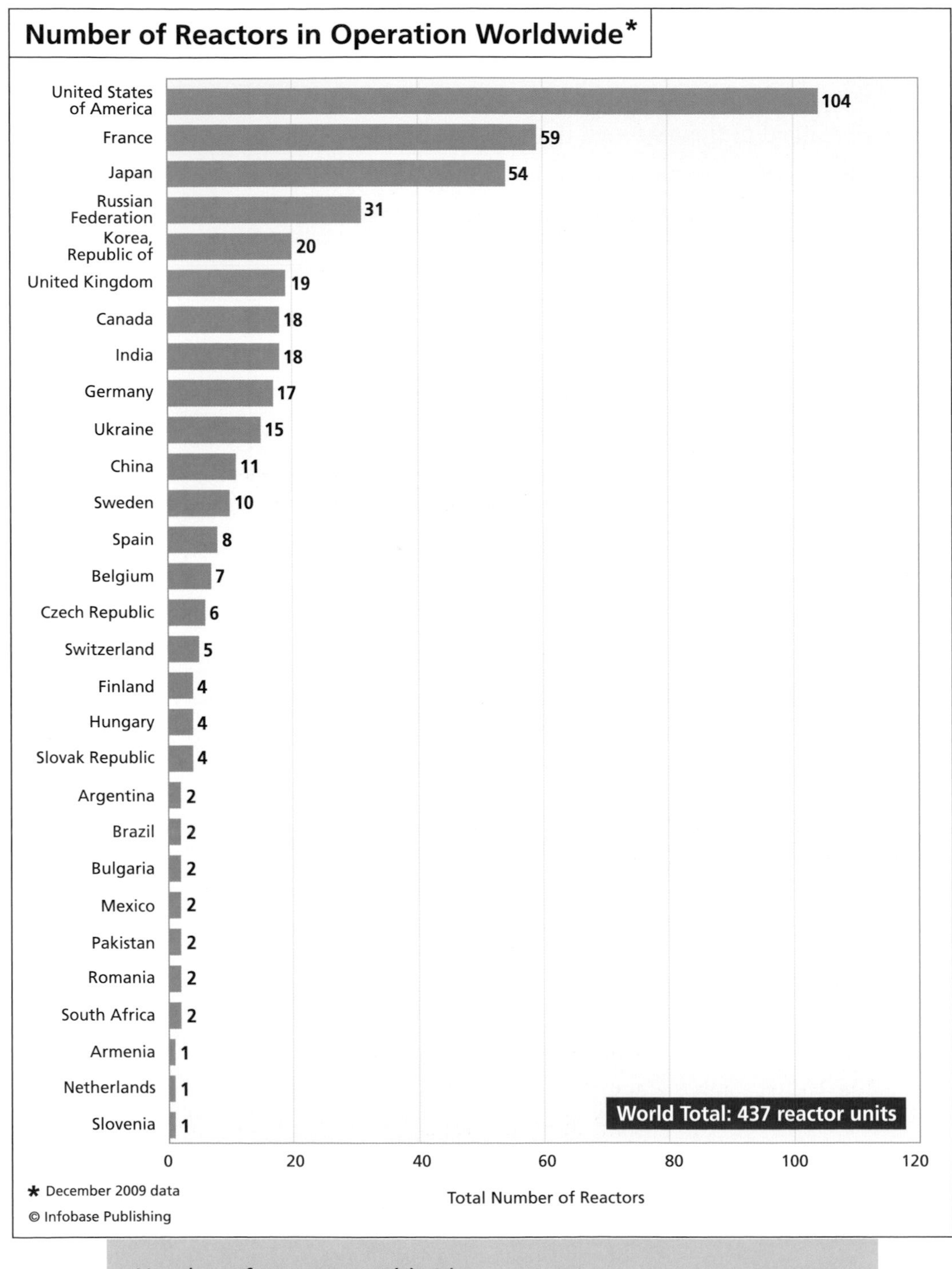

Number of reactors worldwide

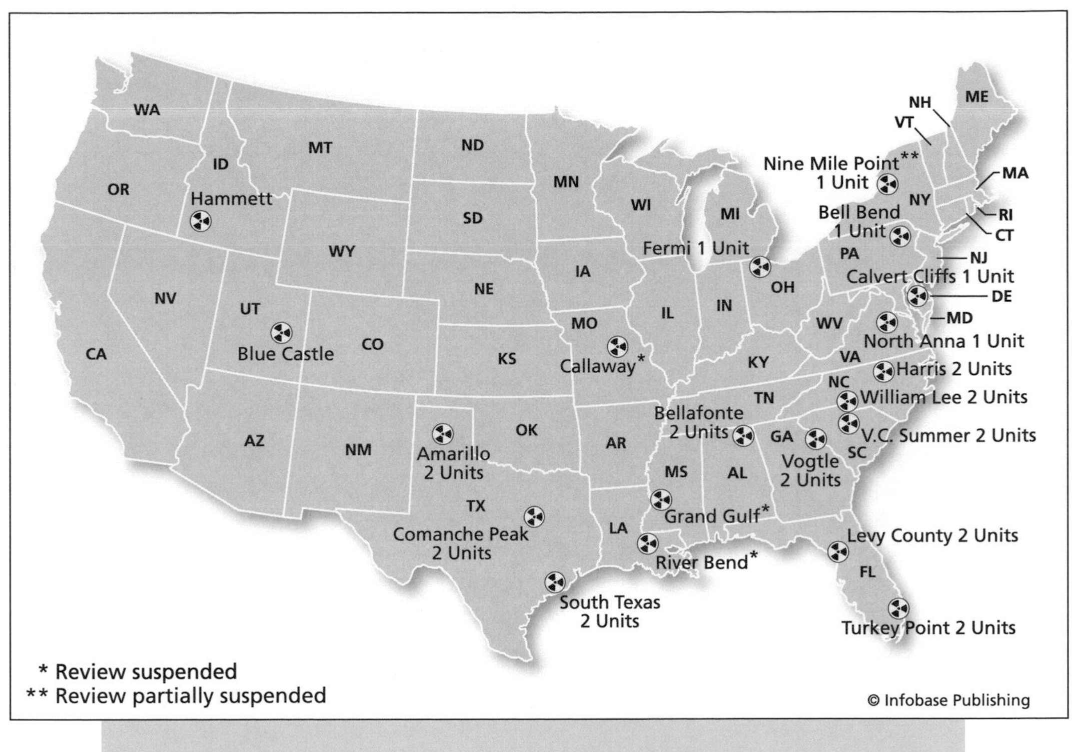

A January 2010 map from the U.S. Nuclear Regulatory Commission that shows the projected locations of proposed new reactors.

DEPLETED URANIUM: A DEBATABLE USE FOR WARTIME AMMUNITION

The process of enriching natural uranium for use in fission reactors (see preceding section) leaves a balance of material with a reduced content of uranium 235. This substance is called depleted uranium (DU)—technically defined to be uranium containing less than 0.711 percent U-235—of which huge stockpiles have accumulated in many countries. The United States alone stores at least 550,000 tons (500,000 metric tons) in the form of solid uranium hexafluoride (DUF_6) in steel containers adjacent to uranium separation facilities. As long as the material remains in the containers, it can do no harm. It must not, however, be allowed to come into contact with water vapor in the air, so the containment vessels must be monitored for corrosion and leaks. Reaction with water will produce the uranium salt *uranyl fluoride* (UO_2F_2), which is

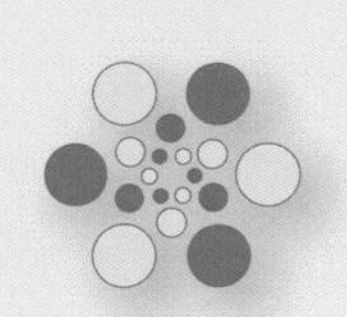

URANIUM AND HUMAN HEALTH

Uranium in its natural environs—embedded beneath Earth's surface—is very rarely a risk to humans or animals (although it is indirectly the source of radon that may be an issue in some homes). It becomes a substantial problem, however, when it is removed from those environs and transported aboveground. The risk to health comes from the alpha particles and electrons emitted during the spontaneous decay of uranium and its daughters. (See the uranium decay chain on page 57.) These particles carry electric charge. Any traveling charged particle can be classified as *ionizing radiation,* which means that it can remove an electron from an atom, molecule, or cell. This is a *carcinogenic* process in the human body, damaging DNA even when the radiation does not invade the nucleus of the cell.

Mine workers are especially at risk for inhalation of radioactive particles, as are people and animals that live downwind of open-pit uranium mines and may, therefore, inhale radioactive dust carried on the breeze. This radioactive dust can also settle on water supplies, but a more common source of water contamination is from rainfall seeping through piles of mine *tailings* into *aquifers* or, more catastrophically, by flooding owing to failure of a dam holding back radioactive sludge. Notable occurrences of the latter include a 2004 dam failure in France as a result of heavy rain, leakage from the Olympic Dam in Australia in 1994, and a large spill in New Mexico in 1979 from sediment settling under a dam. The New Mexico event spilled hundreds of thousands of cubic meters of radioactive water, poisoning sediments up to 70 miles (112.7 km) downstream.

Mountains of tailings from decommissioned uranium mines continue to threaten the environment, one example being the 16-million-ton waste pile near Moab, Utah, left by a mine that closed in 1984. The waste is in such close proximity to the Colorado River that approximately 100,000 gallons (378,541 l) of contaminated water per day seeps into the river—a source of drinking water and irrigation for nearly 10 percent of the pop-

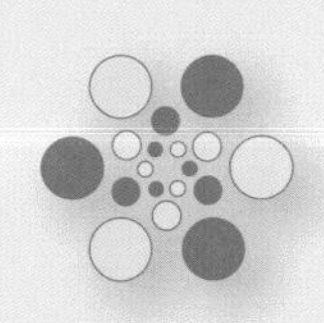

ulation of the United States. This situation is severe enough that in 2009 the U.S. Department of Energy began transferring the waste 30 miles north, a process that will most likely take a decade to complete.

The Colorado Plateau (see map below), which includes the Moab site, has historically been home to hundreds of uranium mining endeavors. This is Navajo country, and the Navajo people, as miners and the primary residents of the area, have long suffered the deleterious health effects—such as cancer, kidney

(continues)

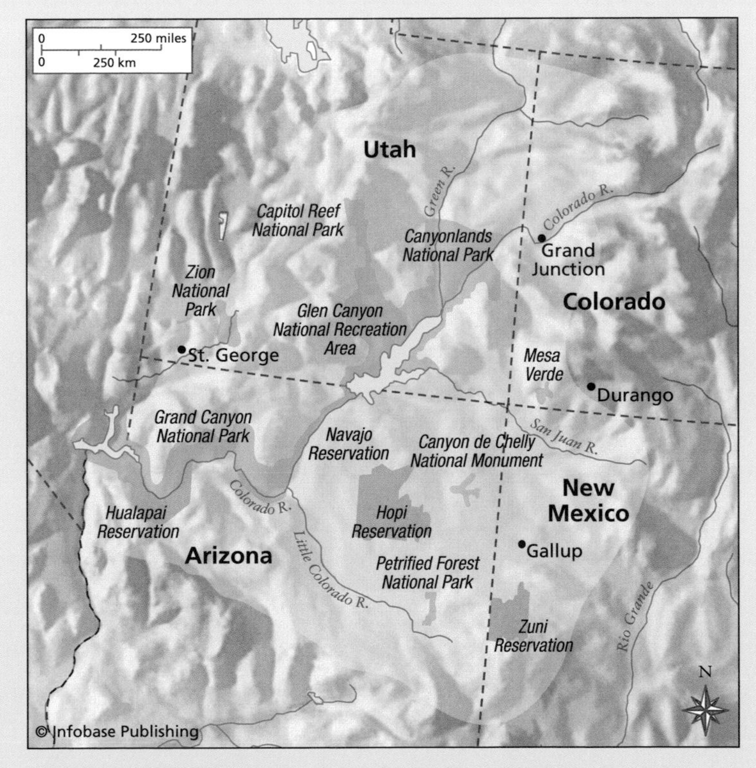

The Colorado Plateau has historically been home to hundreds of uranium mining endeavors.

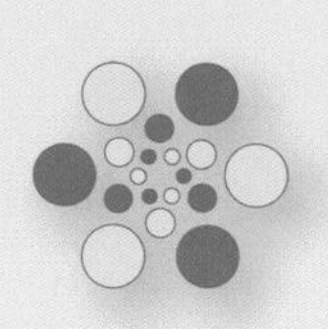

(continued)

disease, and birth defects—of the commercial quest for uranium. Navajo Nation lawyers are still fighting for victim compensation and for cleanup of the radioactive waste.

The U.S. Nuclear Regulatory Commission (NRC) now regulates all aspects of uranium mining from the initial plan to final environmental recovery, but much work remains to be done on the Colorado Plateau and elsewhere. The probable commissioning of new nuclear reactors, however, will increase demand for uranium in the near future. It will be a challenge for the NRC to oversee all new mines and old waste sites in a manner that safeguards the population.

highly soluble and could naturally end up in groundwater. Almost any heavy metal is toxic when ingested or absorbed by humans or animals, and DU is exceptionally heavy, being composed mainly of uranium 238—the heaviest naturally occurring element. Chronic exposure to DU can have negative health effects, which may include an increased risk of leukemia, genetic and reproductive disorders, or neurological problems.

Questions have been raised about such health effects in relation to the use of DU in ammunition and armor-plating. Depleted uranium is denser than lead and has other properties that make it a material of choice for projectiles and shielding against projectiles. Its use in this field dates back to the 1970s, when the Soviet Union developed stronger armor plating for its tanks. One of the properties for which DU is unique is its self-sharpening action as it bores through another material—as opposed to lead projectiles, which become duller, and therefore less penetrating as they pass through a tank wall, for example. The U.S. Air Force now uses depleted uranium ammunition primarily in the aptly named "Tankbuster" gun in the A-10 Thunderbolt jet, for yet another reason: Depleted uranium has a tendency to spontaneously ignite. Tanks targeted by this weapon are not only vulnerable to its ability to penetrate armor plating, but also to the probability that it will ignite explosive fuel and ammunition within.

Depleted uranium weapons were used in the 1991 Gulf War and the 1999 NATO bombing of Serbia. It is estimated that in 2003 munitions containing more than 1,100 tons (1,000 metric tons) of DU were used to bombard Iraqi cities during a three-week period. During impact and combustion, these weapons emitted uranium 238 as an *aerosol.* While quite localized, and therefore of minimal concern to the population of the country as a whole, those citizens living in an area that was subjected to such a bombardment are at risk for day-to-day exposure.

Another concern among the general public relates to fear about uranium's radioactive properties. Depleted uranium presents very little threat in that regard. Its principal danger lies in its toxic heavy metal nature, which brings the question back to the vast quantities stored in the United States. Future applications look to reenriching the stored DU for use in nuclear reactors, which may prove to be an important source of uranium in times when energy needs are high while the costs of mining uranium ore are rising.

Depleted uranium is used in modern military ammunition because it is stronger and more penetrating than lead. *(Visuals Unlimited)*

TECHNOLOGY AND CURRENT USES OF URANIUM

While uranium has been used in the past to give a particular sheen to ceramics and glasses for everyday use, that practice was discontinued decades ago owing to the risks of radioactive exposure. Nowadays, there are only a few applications, though they are important ones.

First and foremost is the application of uranium in nuclear power, which provides electricity to millions of people worldwide. For this use, the most common isotope, uranium 238, must be converted to a more fissionable isotope, uranium 235. Slow neutrons are then used to induce the fission of U-235 that provides heat to turn the turbines that supply *alternating current* to the power companies. Uranium is also used in nuclear weapons, which are covered in chapter 4.

Uranium collisions at the Relativistic Heavy Ion Collider at Brookhaven National Laboratory in New York may give information about conditions very early in the formation of the universe. *(The Relativistic Heavy Ion Collider at Brookhaven National Laboratory)*

In research, collisions of uranium atoms undertaken at the Relativistic Heavy Ion Collider at Brookhaven National Laboratory in New York can simulate hot, dense nuclear matter and assist scientists studying conditions shortly after the big bang.

A method for determining the age of objects up to 400,000 years old is based on the ratio between the amount of thorium 230 and uranium 234 present in the object in question and can determine the age of materials that spent their lifetimes in seawater or other underground aquifers containing uranium. Many cave stalactites and stalagmites and marine organisms, such as ocean corals and fish fossils, are successfully dated using this method. Even the length of time volcanic particles have spent in the atmosphere can be determined by dating of raindrops using this method.

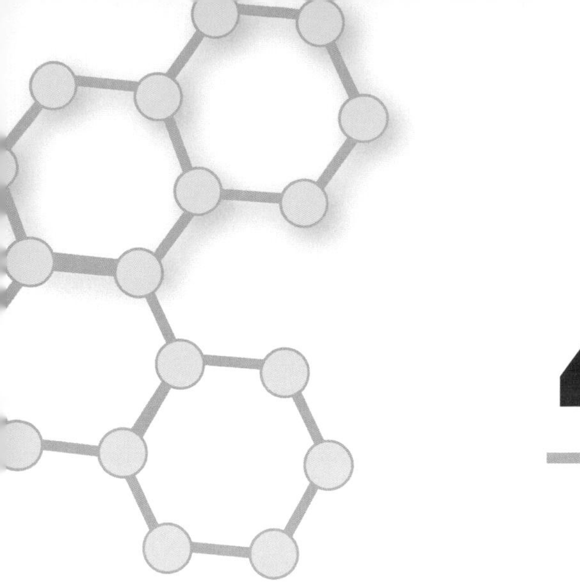

4

The Transuranium Elements

Throughout history, one goal of the alchemists and subsequently of modern chemists and physicists has been the discovery of new elements. Ninety of the first 92 elements occur naturally on Earth. Technetium and promethium are the two exceptions; these elements have to be produced artificially since both elements consist only of radioactive isotopes with relatively short half-lives. After element 92, uranium, all of the elements have had to be produced artificially. (Very minute amounts—in concentrations of parts per trillion or less—of a few elements like plutonium are produced by nuclear processes in the ground, but these quantities will be ignored because they are essentially unrecoverable.) As of the writing of this book, chemists and physicists have discovered a total of 118 elements. Beginning in 1940 with the synthesis of element 93 (neptunium), 25 elements heavier than uranium—the so-called *transuranium elements*—have been produced at laboratories in the United States, Germany, and Russia. The transuranium elements represent 30 percent of the known elements, a significant extension of the periodic table beyond the elements found in nature.

To a large extent, the modern search for elements beyond uranium has been driven primarily by the quest of scientists to discover something new—in this case, to extend the periodic table. However, the discoveries of new elements are more significant than just adding them to the table. The orderly arrangement of elements in the periodic table into vertical groups, or families, possessing similar chemical and physical properties is the reason the periodic table is so useful to scientists. The trends in known elements that have been observed can be used to predict the existence and properties of elements yet to be discovered. But suppose that what appears to be an orderly arrangement exists only for the lightest elements? What if something significantly different happens to the trends in chemical and physical properties as successively heavier elements are discovered? It has been particularly gratifying to scientists to discover that the properties of the transuranium elements have lent strong validation to scientific theories about the most fundamental nature of matter.

An additional motivation for the search for new elements is the possibility that new substances may be discovered that have practical applications in science and technology. The fissionable properties of plutonium are well known, and tons of plutonium have been produced worldwide, especially during the cold war. The next element, americium, has found use in smoke detectors. Unfortunately, the half-lives of the elements following americium are so short that insubstantial quantities of these elements have ever been produced. Theory suggests, however, that there could be elements heavier than any that have been made so far that would have significantly longer half-lives—hundreds, maybe even thousands of years. Should elements of that nature ever be found, they could possibly possess useful properties that cannot be predicted.

This chapter treats the discovery, chemistry, and uses of the first several transuranium elements—neptunium (Np), plutonium (Pu), americium (Am), curium (Cm), berkelium (Bk), californium (Cf), einsteinium (Es), fermium (Fm), mendelevium (Md), nobelium (No), and lawrencium (Lr).

THE BASICS OF THE TRANSURANIUM ELEMENTS

The elements heavier than uranium do not occur naturally. The tables show the longest-lived isotopes of each element. Since transuranium

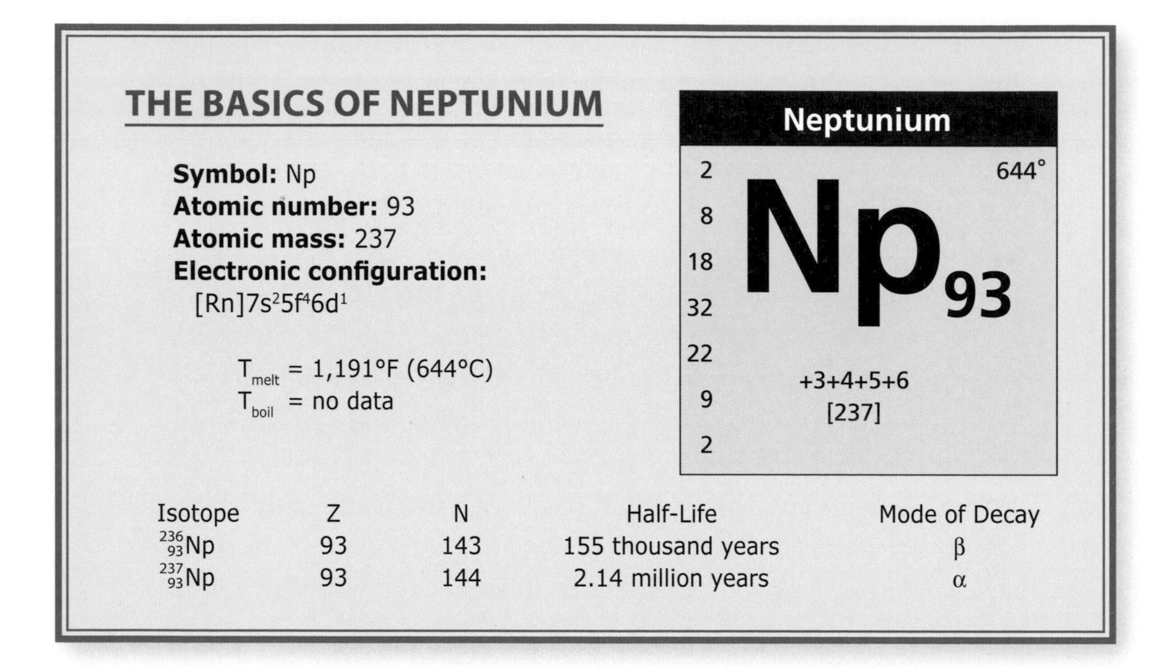

THE BASICS OF NEPTUNIUM

Symbol: Np
Atomic number: 93
Atomic mass: 237
Electronic configuration:
$[Rn]7s^2 5f^4 6d^1$

T_{melt} = 1,191°F (644°C)
T_{boil} = no data

Isotope	Z	N	Half-Life	Mode of Decay
$^{236}_{93}Np$	93	143	155 thousand years	β
$^{237}_{93}Np$	93	144	2.14 million years	α

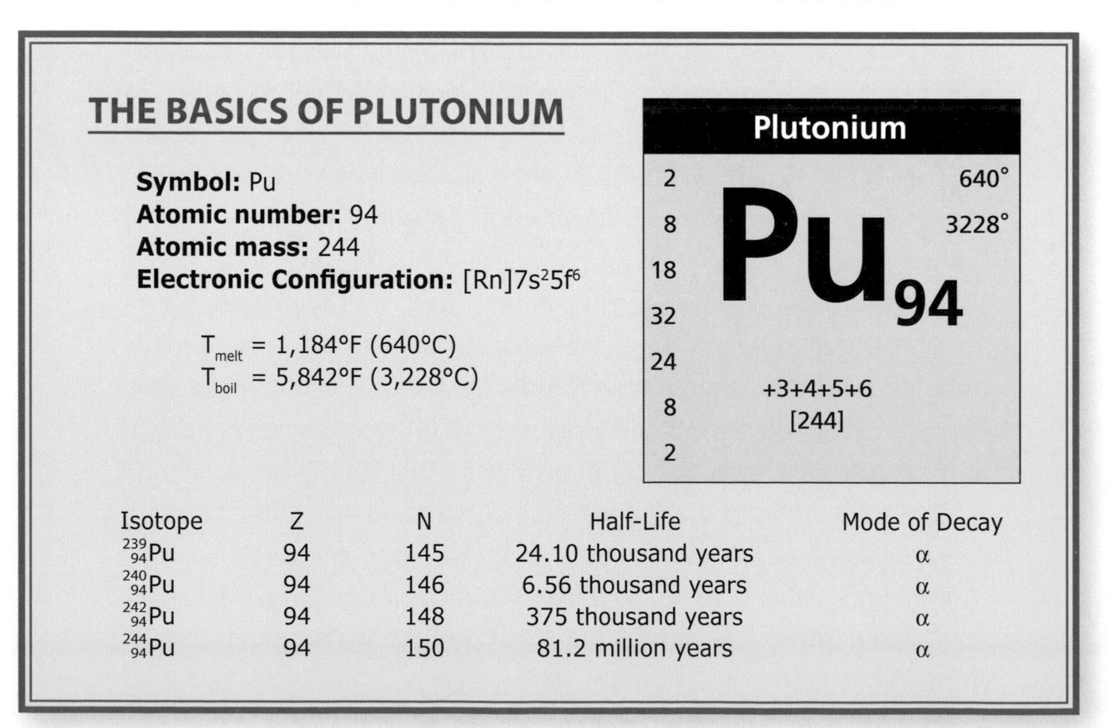

THE BASICS OF PLUTONIUM

Symbol: Pu
Atomic number: 94
Atomic mass: 244
Electronic Configuration: $[Rn]7s^2 5f^6$

T_{melt} = 1,184°F (640°C)
T_{boil} = 5,842°F (3,228°C)

Isotope	Z	N	Half-Life	Mode of Decay
$^{239}_{94}Pu$	94	145	24.10 thousand years	α
$^{240}_{94}Pu$	94	146	6.56 thousand years	α
$^{242}_{94}Pu$	94	148	375 thousand years	α
$^{244}_{94}Pu$	94	150	81.2 million years	α

elements have no natural abundances, the atomic mass listed for an element is the mass number of the longest-lived isotope of that element.

(continues on page 79)

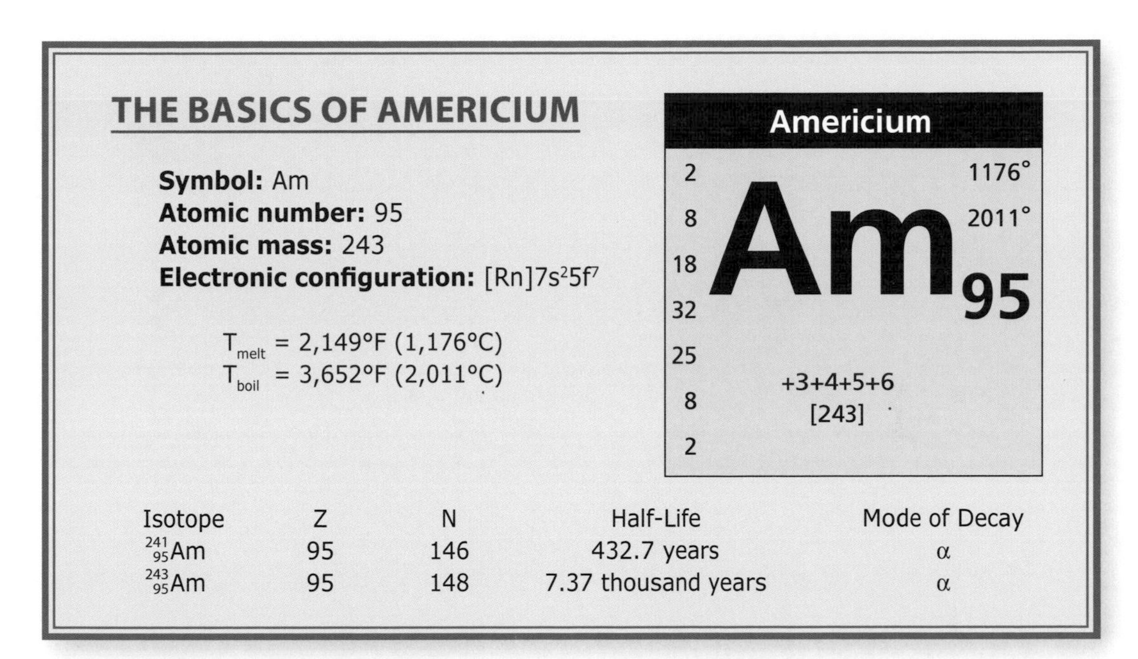

THE BASICS OF AMERICIUM

Symbol: Am
Atomic number: 95
Atomic mass: 243
Electronic configuration: [Rn]$7s^2 5f^7$

T_{melt} = 2,149°F (1,176°C)
T_{boil} = 3,652°F (2,011°C)

Isotope	Z	N	Half-Life	Mode of Decay
$^{241}_{95}Am$	95	146	432.7 years	α
$^{243}_{95}Am$	95	148	7.37 thousand years	α

THE BASICS OF CURIUM

Symbol: Cm
Atomic number: 96
Atomic mass: 247
Electronic configuration:
[Rn]$7s^2 5f^7 6d^1$

T_{melt} = 2,453°F (1,345°C)
T_{boil} = no data

Curium
2
8
18
32
25
9
2
1345°
Cm
96
+3
[247]

Isotope	Z	N	Half-Life	Mode of Decay
$^{245}_{96}Cm$	96	149	8.48 thousand years	α
$^{246}_{96}Cm$	96	150	4.76 thousand years	α
$^{247}_{96}Cm$	96	151	15.6 million years	α
$^{248}_{96}Cm$	96	152	348 thousand years	α
$^{250}_{96}Cm$	96	154	9.7 thousand years	α, sf

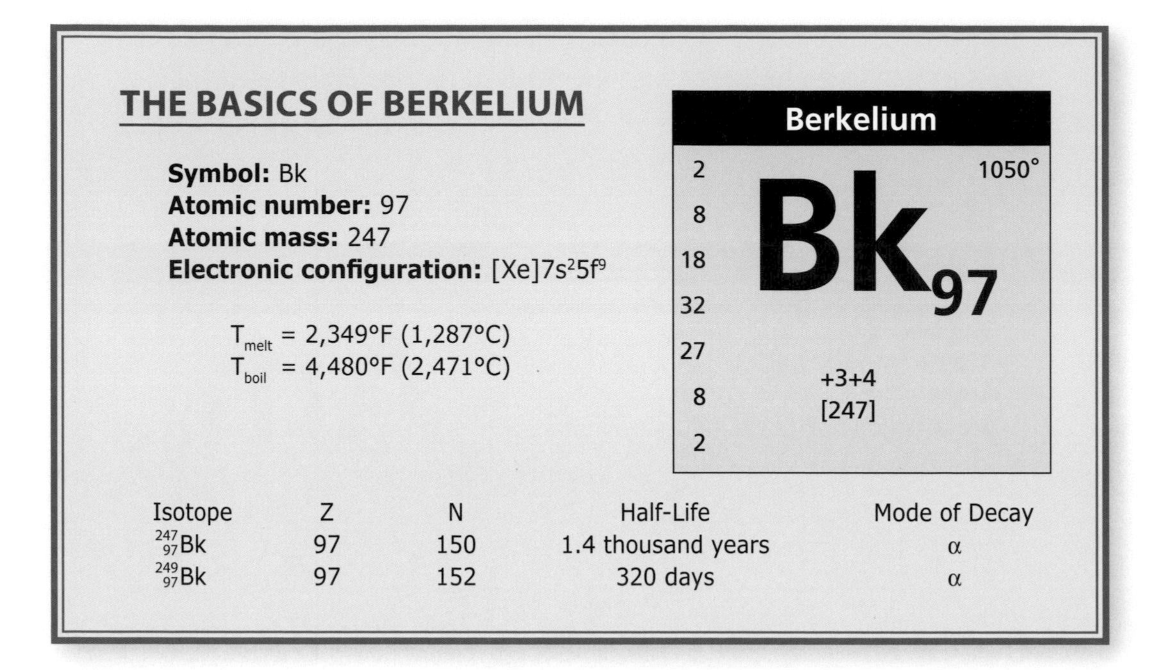

THE BASICS OF BERKELIUM

Symbol: Bk
Atomic number: 97
Atomic mass: 247
Electronic configuration: $[Xe]7s^25f^9$

T_{melt} = 2,349°F (1,287°C)
T_{boil} = 4,480°F (2,471°C)

Isotope	Z	N	Half-Life	Mode of Decay
$^{247}_{97}Bk$	97	150	1.4 thousand years	α
$^{249}_{97}Bk$	97	152	320 days	α

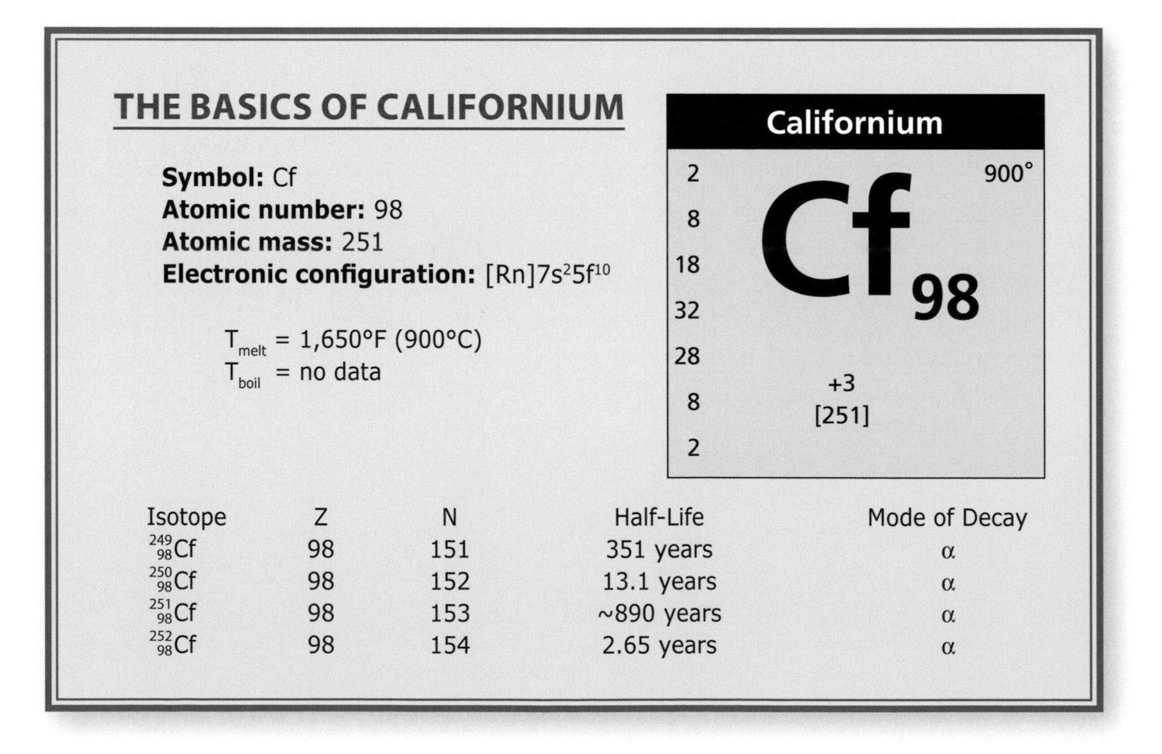

THE BASICS OF CALIFORNIUM

Symbol: Cf
Atomic number: 98
Atomic mass: 251
Electronic configuration: $[Rn]7s^25f^{10}$

T_{melt} = 1,650°F (900°C)
T_{boil} = no data

Isotope	Z	N	Half-Life	Mode of Decay
$^{249}_{98}Cf$	98	151	351 years	α
$^{250}_{98}Cf$	98	152	13.1 years	α
$^{251}_{98}Cf$	98	153	~890 years	α
$^{252}_{98}Cf$	98	154	2.65 years	α

THE BASICS OF EINSTEINIUM

Symbol: Es
Atomic number: 99
Atomic mass: 252
Electronic configuration: [Rn]$7s^25f^{11}$

T_{melt} = ~1,580°F (~860°C)
T_{boil} = no data

Einsteinium	
2	860°
8	Es
18	99
32	
29	+3
8	[252]
2	

Isotope	Z	N	Half-Life	Mode of Decay
$^{252}_{99}Es$	99	153	1.29 years	α
$^{254}_{99}Es$	99	155	276 days	α
$^{255}_{99}Es$	99	156	40 days	β

THE BASICS OF FERMIUM

Symbol: Fm
Atomic number: 100
Atomic mass: 257
Electronic configuration: [Xe]$5f^{12}7s^2$

T_{melt} = 2,781°F (1,527°C)
T_{boil} = no data

Fermium	
2	1527°
8	Fm
18	100
32	
30	+3
8	[257]
2	

Isotope	Z	N	Half-Life	Mode of Decay
$^{252}_{100}Fm$	100	152	1.058 days	α
$^{253}_{100}Fm$	100	153	3.0 days	α
$^{257}_{100}Fm$	100	157	100.5 days	α

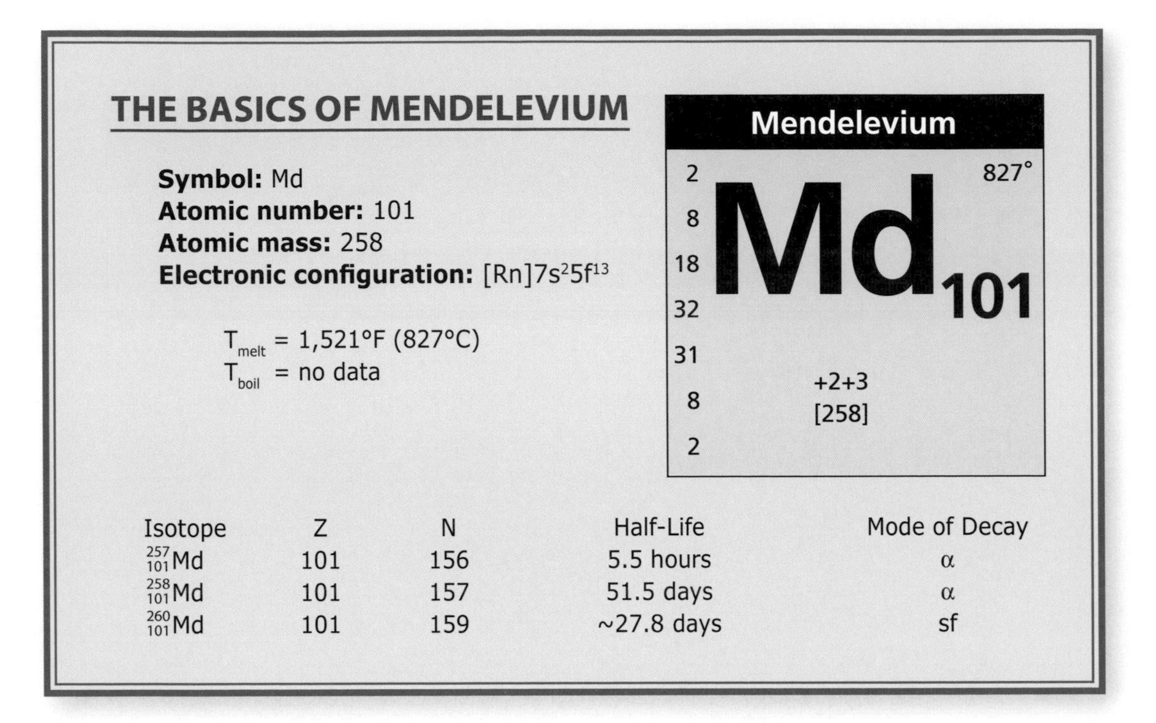

THE BASICS OF MENDELEVIUM

Symbol: Md
Atomic number: 101
Atomic mass: 258
Electronic configuration: $[Rn]7s^25f^{13}$

T_{melt} = 1,521°F (827°C)
T_{boil} = no data

Mendelevium
2 8 18 32 31 8 2
827°
Md_{101}
+2+3
[258]

Isotope	Z	N	Half-Life	Mode of Decay
$^{257}_{101}Md$	101	156	5.5 hours	α
$^{258}_{101}Md$	101	157	51.5 days	α
$^{260}_{101}Md$	101	159	~27.8 days	sf

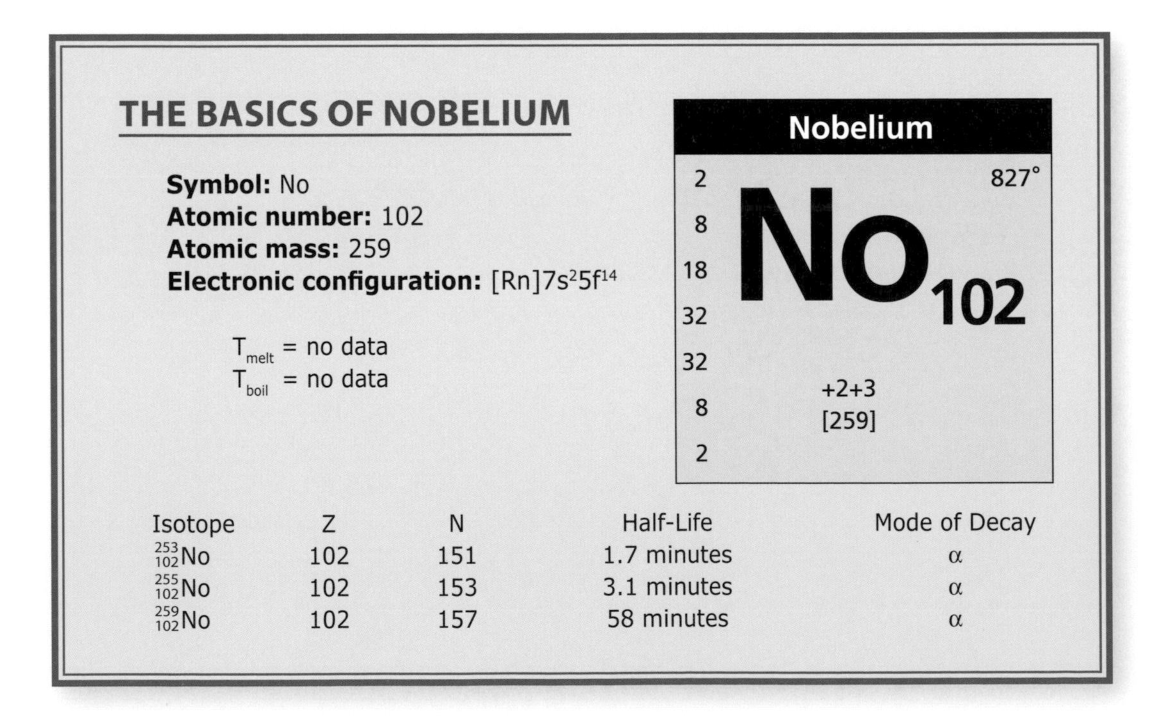

THE BASICS OF NOBELIUM

Symbol: No
Atomic number: 102
Atomic mass: 259
Electronic configuration: $[Rn]7s^25f^{14}$

T_{melt} = no data
T_{boil} = no data

Nobelium
2 8 18 32 32 8 2
827°
No_{102}
+2+3
[259]

Isotope	Z	N	Half-Life	Mode of Decay
$^{253}_{102}No$	102	151	1.7 minutes	α
$^{255}_{102}No$	102	153	3.1 minutes	α
$^{259}_{102}No$	102	157	58 minutes	α

THE BASICS OF LAWRENCIUM

Symbol: Lr
Atomic number: 103
Atomic mass: 262
Electronic configuration:
[Rn]$7s^2 5f^{14} 6d^1$

T_{melt} = no data
T_{boil} = no data

Lawrencium
2
8
18
32
32
9
2
1627°
Lr
103
+3
[262]

Isotope	Z	N	Half-Life	Mode of Decay
$^{251}_{103}Lr$	103	148	39 minutes	sf
$^{260}_{103}Lr$	103	157	3 minutes	α
$^{261}_{103}Lr$	103	158	40 minutes	sf
$^{262}_{103}Lr$	103	159	3.6 hours	sf

(continued from page 75)

DISCOVERY AND NAMING OF TRANSURANIUM ELEMENTS

The following accounts of the discoveries of elements 93–103 relate events that occurred mostly at the Lawrence Berkeley Laboratory (at that time the Lawrence Radiation Laboratory), which sits on the hillside overlooking the University of California, Berkeley, campus. Some events occurred elsewhere (for example, at the University of Chicago in Illinois, at Oak Ridge National Laboratory in Tennessee, or at the Hanford Laboratory in the state of Washington), in part because of wartime activities. Glenn Seaborg (1912–99) of the University of California at Berkeley discovered plutonium and contributed to the discoveries of most of the other transuranium elements (see sidebar). A key scientist whose work has been seminal in the discovery of transuranium elements has been Seaborg's longtime colleague Albert Ghiorso (1915–), who at the time this is being written is still an active researcher. In fact, Ghiorso holds the Guinness World Record for having discovered the most elements (12).

(continues on page 82)

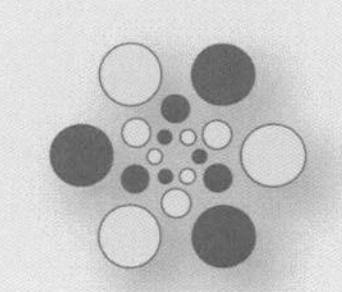

GLENN THEODORE SEABORG

The nuclear chemist whose name is most associated with the discoveries of the transuranium elements is Glenn Theodore Seaborg (1912–99). Seaborg was born to Swedish immigrants in the Upper Peninsula of Michigan; his father worked in the iron mines. Later, Seaborg's family moved to Southern California, where Seaborg pursued his undergraduate studies at the University of California, Los Angeles. After graduating from UCLA, he moved to the University of California, Berkeley, and earned his Ph.D. in chemistry under the leadership of the dean of the College of Chemistry, Gilbert Newton Lewis. Seaborg remained at Berkeley and became a senior scientist at the Lawrence Radiation Laboratory (now the Lawrence Berkeley Laboratory, named for the Berkeley physicist and Nobel laureate Ernest Orlando Lawrence), where his research resulted in the discoveries of 10 new elements—plutonium, americium, curium, berkelium, californium, einsteinium, fermium, mendelevium, nobelium, and seaborgium. Beginning at the age of 14, Seaborg recorded each day's events nightly in his diary. His collection of diaries is now a rich source of material for his biographers.

From 1958 to 1961, Seaborg served as chancellor of the Berkeley campus. Except for leaves of absence spent conducting war-related research at the University of Chicago during World War II and serving as the director of the Atomic Energy Commission from 1961 to 1971, Seaborg remained at Berkeley for the rest of his life. Seaborg served as science adviser to every president of the United States from Harry Truman to George H. Bush. Seaborg served terms as president of both the American Chemical Society and the American Association for the Advancement of Science. In 1979, the American Chemical Society recognized his lifetime of achievement with its most prestigious award, the Priestley Medal (named for Joseph Priestley, the discover of oxygen).

Seaborg was passionate about science and science education. After he returned to the chemistry department at Berkeley

The nuclear chemist whose name is most associated with the discoveries of the transuranium elements is Glenn Theodore Seaborg. *(Fritz Goro/Time & Life Pictures/Getty Images)*

in 1971, Seaborg could be found on the second floor of Latimer Hall, teaching sections of freshman chemistry laboratories and mentoring students in his fourth-floor office. He played an active role in the establishment and administration of the Lawrence Hall of Science, a center for public science education located on the hillside overlooking the Berkeley campus. He was never too busy to spend time chatting with teachers and students. Just imagine how exciting it must have been for his grandchildren when Seaborg would visit their elementary school classrooms and perform chemistry demonstrations for their classmates!

In 1951, Seaborg shared the Nobel Prize in chemistry with the director of the Lawrence Radiation Laboratory, Edwin McMillan. Because Seaborg's parents were Swedish immigrants,

(continues)

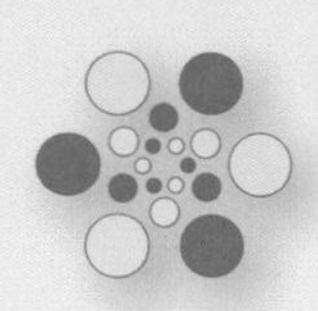

(continued)

Swedish was spoken in Seaborg's home when he was growing up. Therefore, when presented with the Nobel Prize in Stockholm, Seaborg delivered his acceptance speech in Swedish, much to the delight of the king of Sweden.

During his lifetime, Seaborg was instrumental in the discovery of more than 100 new isotopes. Particularly gratifying to him was the discovery of radioactive isotopes of iodine, I-131, and cobalt, Co-60. Both isotopes are used in hospitals every day to treat patients with life-threatening diseases. The life of Seaborg's own mother was saved from thyroid disease because of treatment with iodine-131.

Seaborg had the distinction of having been still very much an active researcher at the time an element was named for him. His collaborators in the discovery of element 106 voted to recommend to the International Union for Pure and Applied Chemistry that element 106 be named *seaborgium* in his honor. At first, the IUPAC balked at naming an element for someone who was still living but eventually voted to affirm the discoverers' choice.

Glenn Seaborg was married to the former Helen Griggs, Ernest Lawrence's secretary. Together, they had seven children. Seaborg was a devoted husband and father. Upon the announcement of the naming of element 106 seaborgium at a meeting of the American Chemical Society, a member of the audience asked Seaborg what he felt was his greatest discovery. Without hesitation, Seaborg replied, "My wife, Helen."

Glenn Seaborg remained active in scientific research right up to his participation at the August 1998 national meeting of the American Chemical Society. Seaborg suffered a stroke at the meeting. After several months of convalescence at his home in Lafayette, California, Seaborg died in February 1999. He was preceded in death by one son. His wife died in 2006.

(continued from page 79)

During the Middle Ages, it had been the dream of the alchemists to convert base metals such as lead into gold. All of their attempts failed because the conversion (or *transmutation*) of one element into another element requires a nuclear reaction. The alchemists had no knowledge of atoms, let alone the composition of atoms. Performing nuclear transmutations was simply beyond their capability.

That situation changed with the discoveries of elementary particles—protons and electrons—and of radioactivity. In England, Ernest Rutherford (1871–1937) was the first person who successfully changed one element into another element. Rutherford was already famous for his work on radioactivity and for his discovery of atomic nuclei. (In 1908, Rutherford received the Nobel Prize in chemistry for his work on radioactivity.) Neutrons remained undiscovered at that time, so Rutherford was unaware of them. Nevertheless, he proposed a *nuclear model* of the atom that had the protons (and some of the electrons) in a very tiny center that he called the *nucleus,* surrounded by the remaining electrons at a considerable distance from the nucleus. Rutherford recognized that the identity of an atom is determined by the number of protons in the atom's nucleus. Therefore, changing the number of protons would change the atom's identity.

The physical problem that had to be overcome is that protons all carry positive charges; therefore, they repel one another. The force of repulsion increases dramatically as the protons approach one another. However, if they manage to get close enough to one another, they can combine to make a new nucleus. (They do so because at close enough distances, the more powerful nuclear force of attraction takes over, although Rutherford would not have understood nuclear forces at the time.) In 1919, Rutherford achieved the first transmutation of an element. By bombarding nitrogen nuclei with helium nuclei (also called *alpha particles*), he was able to convert nitrogen atoms into oxygen atoms using the following reaction:

$$^{14}_{7}N + ^{4}_{2}He \rightarrow ^{17}_{8}O + ^{1}_{1}H,$$

in which subscripts indicate charges on species, and superscripts indicate mass numbers. Charge must be conserved; hence, the sum of the charges on the left-hand side equals 9, as does the sum of the charges on the right-hand side. In the case of an element, the subscript also designates the atomic number of the element, which corresponds to the number of protons in that element's atoms. (If the species is not an element, then the subscript is simply the charge on that species.) The mass number of a particle is the sum of its protons and neutrons. That sum also must be conserved; hence, the sum of the superscripts on the left-hand side equals 18, as does the sum of the superscripts on the right-hand side. For several years thereafter, bombarding atomic nuclei with alpha particles was the most common method for transmuting elements.

That situation changed dramatically in 1932 with the discovery of the neutron by the English physicist James Chadwick (1891–1974), who worked at the Cavendish Laboratory under Rutherford's leadership. (Chadwick received the Nobel Prize in physics in 1935 for this discovery.) Because neutrons are electrically neutral, they are not repelled by atomic nuclei. Scientists quickly realized that their neutrality made neutrons excellent projectiles to fire at nuclei.

Generally speaking, the lighter transuranium elements have been made by bombarding nuclei of uranium and other elements with light particles—neutrons, deuterons, or alpha particles. Heavier isotopes of the parent element are produced, which then undergo beta decay resulting in the formation of the next element to the right in the periodic table. This technique works only up to a point. Eventually, so little of the heaviest possible target material is available for bombardment that lighter targets and heavier projectiles—carbon, oxygen, or even heavier nuclei—have to be used. The choice of targets and projectiles is determined by the requirement that the charges (atomic numbers in the cases of elements) of the two species must add up to the charge (atomic number) on the nucleus of the desired product.

For example, in the synthesis of element 93 (neptunium), the following reaction was used:

$$^{238}_{92}\mathrm{U} + ^{1}_{0}\mathrm{n} \rightarrow ^{239}_{92}\mathrm{U} \rightarrow ^{239}_{93}\mathrm{Np} + ^{0}_{-1}\beta.$$

This equation shows that nuclei of U-238 were bombarded with neutrons ($^{1}_{0}\mathrm{n}$ is the symbol for a neutron) to produce unstable U-239 nuclei.

The U-239 nuclei underwent beta decay ($^{0}_{-1}\beta$ is the symbol for a beta particle) resulting in Np-239.

Early Attempts

Various groups performed experiments in which they added neutrons to uranium atoms in expectation of creating elements heavier than uranium (transuranium elements). One group was led by the German chemist Otto Hahn (1879–1968) and the Austrian physicist Lise Meitner (1878–1968) in Berlin. They failed to produce transuranium elements. More important, however, in 1938, they discovered *nuclear fission.* (Hahn was awarded the 1944 Nobel Prize in chemistry for this discovery. Meitner did not share the Nobel Prize; element 109, however, has been named in her honor.) An Italian group was led by the physicists Enrico Fermi (1901–54) and Emilio Segrè (1905–89). Fermi's group also failed to produce transuranium elements. His group also failed to recognize fission, leaving that discovery for Hahn and Meitner.

Although their search for transuranium elements was unsuccessful, both Fermi and Segrè achieved a number of successes that included important contributions to the Manhattan (atom bomb) Project and separate Nobel Prizes in physics for each of them. Fermi gave the first complete explanation of beta decay, built the first nuclear reactor that demonstrated the first controlled fission chain reaction, and discovered a set of statistical laws called *Fermi-Dirac statistics* that govern the behavior of certain elementary particles now called *fermions.* Fermi received the 1938 Nobel Prize in physics and was honored at the time of his death with the naming of element 100 *(fermium)* in his honor. In addition to all of Fermi's scientific discoveries, he was also highly esteemed as a professor of physics first at Columbia University and later at the University of Chicago.

Segrè was a codiscoverer of two elements—technetium (number 43) and astatine (number 85)—and of the isotope of plutonium (Pu-239) that was discovered to be fissionable and became the isotope of choice in nuclear weapons. In collaboration with the American physicist Owen Chamberlain (1920–2006), Segrè discovered the *antiproton,* an achievement that won Segrè and Chamberlain the 1959 Nobel Prize in physics. From 1938 until his retirement in 1972, Segrè was a highly respected researcher at the Lawrence Radiation Laboratory (later the Lawrence Berkeley Laboratory) and professor of physics at the University of California at Berkeley.

Neptunium

It was at Berkeley that the first transuranium elements were finally discovered. In spring 1940, the physicists Philip H. Abelson and Edwin M. McMillan used the 60-inch *cyclotron* to produce neutrons. (The cyclotron was also used to discover plutonium, curium, berkelium, californium, and mendelevium.) A uranium oxide target was irradiated with the neutrons, resulting in the following reaction:

$$^{238}_{92}U + ^{1}_{0}n \rightarrow ^{239}_{92}U \rightarrow ^{239}_{93}Np + ^{0}_{-1}\beta.$$

Because the planet Neptune follows the planet Uranus, element 93, which follows uranium (element 92), was named *neptunium* and given the symbol *Np*.

In 1942, neptunium 237 was produced by Arthur C. Wahl and Glenn Seaborg. The first weighable sample (10 micrograms) of neptunium (Np-237) as neptunium oxide (NpO_2) was produced at the wartime Metallurgical Laboratory of the University of Chicago (now Argonne National Laboratory) in October 1944. Later, the first sample of metallic neptunium was also produced at the Metallurgical Laboratory.

Neptunium has 23 known isotopes, all of them radioactive. Half-lives range from about two microseconds for Np-225 to 2.14 million years for Np-237. Minuscule quantities of neptunium are produced by nuclear reactions in nature, but neptunium in multikilogram quantities are produced in nuclear reactions as part of the process used to convert uranium into plutonium.

Plutonium

In December 1940, Edwin M. McMillan, Joseph W. Kennedy, and Arthur C. Wahl fired *deuterons* (a form of *heavy hydrogen* containing one proton and one neutron per atom and symbolized $^{2}_{1}H$) at a target of uranium oxide (U_3O_8). Because the characteristics of the product's radioactivity was different than the characteristics of Np-239, they surmised that they had produced a different isotope of neptunium. In fact, what they had made was neptunium 238, as shown by the following reaction:

$$^{238}_{92}U + ^{2}_{1}H \rightarrow ^{238}_{93}Np + 2\,^{1}_{0}n.$$

Neptunium 238 was decaying to plutonium 238 with a half-life of 2.12 days, versus 2.36 days for Np-239. That difference alone was small. More important, Np-238 and Np-239 are both beta emitters. What the scientists observed, however, was an increasing amount of alpha radiation that had to be coming from an entity other than Np-238 or Np-239, most likely a daughter with a higher atomic number. The conclusion was that element 94 had been produced, as shown in the following equation:

$$^{238}_{93}Np \rightarrow ^{238}_{94}Pu + ^{0}_{-1}\beta,$$

where Pu-238 emits alpha particles with a half-life of about 90 years.

By this time, McMillan had left Berkeley for the Massachusetts Institute of Technology. In early 1941, Kennedy, Wahl, and Seaborg set to work trying to prove by chemical analysis that element 94 had in fact been produced. During the night of February 23, 1941, they demonstrated that they had a product that was positively chemically different from neptunium. In March 1942, following McMillan's example in the naming of neptunium, it was decided to name element 94 *plutonium* after Pluto, the next planet after Neptune, and give it the symbol *Pu.* The room in Gilman Hall on the University of California campus where the definitive tests took place is now a National Historic Landmark. The first weighable plutonium compound (PuO_2) was produced in September 1942. It had a mass of 2.77 micrograms.

Later in 1941, Seaborg, Kennedy, Wahl, and Emilio Segrè discovered plutonium 239, which is produced by beta decay of neptunium 239. Segrè discovered that the fission *cross section* of Pu-239 was about 50 percent greater than the cross section of U-235, which means that Pu-239 undergoes fission more readily than U-235 does. Plutonium 239 thus became the material of choice for nuclear weapons because smaller quantities are required for warheads. Also, because Pu-239 has a half-life of 24 thousand years, it can be produced and stockpiled in large quantities.

The discovery of plutonium occurred during World War II, which required that all of the research on it be classified and kept secret until after the end of the war. The public learned of plutonium only when it was revealed that the atom bomb tested on July 16, 1945, at the Trinity Test Site in New Mexico and the bomb dropped on Nagasaki, Japan, on August 9, 1945, both had cores made of plutonium. (The bomb dropped on Hiroshima on August 6 was made of uranium.)

Plutonium has 21 isotopes ranging in mass number from 228 to 247. There are longer-lived isotopes than either Pu-238 or Pu-239. Plutonium 242 has a half-life of 375 thousand years. Plutonium 244 has a very long half-life of 81 million years. These longer-lived isotopes are the ones usually used by chemists to study plutonium's chemistry.

Because plutonium is radioactive and toxic, it is very hazardous to work with, especially if it is inhaled in any form. It is usually studied in only milligram or smaller quantities, and even then with shielding to protect workers from exposure to its radiation.

Americium

Before World War II, actinium, thorium, protactinium, and uranium had been positioned in the periodic table together with the transition elements. Actinium was located at the bottom of Group IIIA under scandium, yttrium, and lanthanum. Thorium was located at the bottom of Group IVA under titanium, zirconium, and hafnium. Protactinium and uranium followed thorium. In 1944, Glenn Seaborg made the bold suggestion that these elements were misplaced. They did not closely resemble the transition metals. Instead, they more closely resembled the lanthanides and should be placed in a row below the lanthanides. Called the *actinide concept,* this was a bold suggestion because it would represent the first major change in the structure of the periodic table since Mendeleev's work in the 19th century. Seaborg's actinide concept proved to be correct and was used as an aid in the discovery of the *transplutonium* elements beginning with americium and curium.

In 1944, while working at the wartime Metallurgical Laboratory in Chicago, Glenn Seaborg, Albert Ghiorso, Ralph A. James, and Leon O. (Tom) Morgan began to look for the next two elements that would follow plutonium in the periodic table. Samples of plutonium were sent to other laboratories around the country where they were irradiated with neutrons or deuterons and the resulting materials analyzed. The first experiments proved to be unsuccessful.

Finally, in July 1944, a plutonium target was bombarded with alpha particles at Berkeley. Curium (element 96) was detected among the products. It had been produced by the following reaction:

$$^{239}_{94}\text{Pu} + {}^{4}_{2}\text{He} \rightarrow {}^{242}_{96}\text{Cm} + {}^{1}_{0}\text{n}.$$

Curium 242 decayed with a half-life of 162.8 days. The first compound of curium, $Cu(OH)_3$, was made in 1947. Today, there are 20 known isotopes of curium with half-lives as short as 51 seconds for Cm-234 to as long as 15.6 million years for Cm-247.

In like manner, plutonium 239 was irradiated with neutrons to produce plutonium 241, which decayed to americium according to the following reactions:

$$^{239}_{94}\text{Pu} + 2\,{}^{1}_{0}\text{n} \rightarrow {}^{241}_{94}\text{Pu} \rightarrow {}^{241}_{95}\text{Am} + {}^{0}_{-1}\beta.$$

Irradiation of americium 241 with neutrons was discovered to also be a pathway to making curium, as shown by the following reactions:

$$^{241}_{95}\text{Am} + {}^{1}_{0}\text{n} \rightarrow {}^{242}_{95}\text{Am} \rightarrow {}^{242}_{96}\text{Cm} + {}^{0}_{-1}\beta.$$

The first compound of americium, $Am(OH)_3$, was made in 1945. Twenty isotopes of americium have been discovered, with half-lives varying from 0.9 minute for Am-232 to 7.37 thousand years for Am-243. In 1964, Glenn Seaborg patented americium and the method for producing it, giving him the distinction of being the only person who has ever held a patent on a chemical element.

The chemistry of both americium and curium was found to be much more similar to their lanthanide *homologues* (europium and gadolinium) than to plutonium. In fact, americium and curium are so similar that it was very difficult to separate them. The scientists who first studied them became so frustrated by their failure to separate them that they somewhat jokingly referred to these elements as “pandemonium” and “delirium”! It was subsequently found that it was much easier to separate these elements using ion exchange methods than by using chemical reactions.

As was the case with the discovery of plutonium, americium and curium were discovered during wartime, so their discovery was not announced to the world until after the war. As it turned out, the announcement of these elements was made in a way probably unique in the history of science. In November 1945, Glenn Seaborg was invited to be a guest on the radio program *Quiz Kids* in Chicago. Information about the elements discovered during the war had just been declassified, and Seaborg was waiting for an upcoming chemistry symposium to make the announcement about elements 95 and 96 (which had not yet been named). One of the children on the radio show, however, asked Seaborg if any other new elements had been discovered besides neptunium and plutonium. Seaborg responded that, yes, two new elements had been made—numbers 95 and 96. The children on the show, and the public listening to the show, thus knew about elements 95 and 96 before the rest of the scientific community did! When Seaborg then said that

Element 96 was named curium in honor of Pierre and Marie Curie, the latter shown here on December 20, 1923, at an unknown location. *(AP Photo)*

these elements did not have names yet, the announcer for *Quiz Kids* invited the listening public to send in their suggestions. Notably, one of the children participating that day was future Nobel laureate James D. Watson, who shared the 1962 Nobel Prize in physiology or medicine for the elucidation of the structure of DNA.

Several suggestions were made for naming element 95, including *sunonium, artifium, artifician, mechanicium, and curium.* In 1946, in recognition that the properties of element 95 are very similar to its lanthanide homologue europium—and europium is named for Europe—element 95 was named *americium* for America. Element 96 is very much like its homologue gadolinium, and gadolinium is named in honor of the chemist Johan Gadolin. Element 96, therefore, was named *curium* in honor of Pierre and Marie Curie.

Berkelium and Californium

The scientists from Berkeley who had worked at the Metallurgical Laboratory in Chicago during World War II eventually returned to Berkeley after the war. In anticipation of trying to synthesize elements 97 and 98, scientists had to produce sufficient quantities of americium and curium to use as the targets. To produce element 97, an americium 241 target was bombarded with alpha particles. In December 1949, Albert Ghiorso, Stanley G. Thompson, and Glenn Seaborg achieved production of element 97 using the following reaction:

$$^{241}_{95}\mathrm{Am} + ^{4}_{2}\mathrm{He} \rightarrow ^{243}_{97}\mathrm{Bk} + 2\,^{1}_{0}\mathrm{n}.$$

The isotope of element 97 that was made has a half-life of 4.5 hours.

In February 1950, Stanley Thompson, Kenneth Street, Albert Ghiorso, and Glenn Seaborg succeeded in producing element 98. They bombarded a curium 242 target with alpha particles as shown in the following reaction:

$$^{242}_{96}\mathrm{Cm} + ^{4}_{2}\mathrm{He} \rightarrow ^{245}_{98}\mathrm{Cf} + ^{1}_{0}\mathrm{n}.$$

In both cases, the scientists used ion-exchange chromatography to separate elements 97 and 98 from the americium and curium targets.

Element 97's lanthanide homologue is terbium, named for the town of Ytterby in Sweden. Therefore, element 97 was named *berkelium* for the city of Berkeley, California, where the university is located. Element 98's homologue is dysprosium, which means "hard to get at." The discoverers of element 98 decided not to try to find an analogous term. Instead, they adopted the name *californium* after both the state and the university. Seaborg defended the name by saying that there is a resemblance to "hard to get at" because California was hard for the Gold Rush 49ers to get to.

In 1958, Burris Cunningham and Stanley Thompson prepared the first berkelium and californium compounds. Today, there are 15 known isotopes of berkelium and 20 isotopes of californium. Half-lives for berkelium range from 2.4 minutes for Bk-238 to 1,400 years for Bk-247. Half-lives for californium range from 21 milliseconds for Cf-238 to 900 years for Cf-251.

Einsteinium and Fermium

The story of the discoveries of new elements takes an unexpected turn at elements 99 and 100. All of the previously transuranium elements were created deliberately as the results of carefully planned experiments. In contrast, the discoveries of elements 99 and 100 were completely unplanned.

The first thermonuclear (hydrogen bomb) explosion conducted by the United States was code-named "Mike" and conducted in the Pacific Basin on November 1, 1952. An airplane flew through the cloud and collected the debris from the blast on filter paper. The filter paper was returned to the United States for investigation. The discovery of elements 99 and 100 in the debris meant that uranium 238 (used to trigger the thermonuclear explosion) had to have acquired a large number of neutrons. The heavy isotopes of uranium that resulted then underwent rapid successions of beta decays until elements 99 and 100 were reached.

To produce element 99, each uranium atom had to have absorbed 15 neutrons to yield uranium 253, which subsequently underwent seven beta decays to reach Es-253. (Recall that during beta decay an element transmutes into the element located to its right in the periodic table, with no change in mass number.) To produce element 100, each uranium atom absorbed 17 neutrons to yield uranium 255, which subsequently underwent eight beta decays to reach Fm-255. The results of the beta decays in each case are shown in the following sequences of events:

$$^{253}_{92}U \rightarrow ^{253}_{93}Np \rightarrow ^{253}_{94}Pu \rightarrow ^{253}_{95}Am \rightarrow ^{253}_{96}Cm \rightarrow ^{253}_{97}Bk \rightarrow ^{253}_{98}Cf \rightarrow ^{253}_{99}Es.$$

$$^{255}_{92}U \rightarrow ^{255}_{93}Np \rightarrow ^{255}_{94}Pu \rightarrow ^{255}_{95}Am \rightarrow ^{255}_{96}Cm \rightarrow ^{255}_{97}Bk \rightarrow ^{255}_{98}Cf \rightarrow ^{255}_{99}Es \rightarrow ^{255}_{100}Fm.$$

It should be noted in each series that the superscripts (mass numbers) do not change. In each successive beta decay, however, the subscript (atomic number) increases by one unit.

Albert Ghiorso made the suggestion to begin a new system of nomenclature, one in which elements would be named after famous scientists. (Previously, Ghiorso was the one who had proposed the name *curium* for element 96.) At the time, two of the world's most famous physicists were Albert Einstein (1879–1955) and Enrico Fermi (1901–54). Fermi was dying of cancer. The announcement to name element 99 *einsteinium* and element 100 *fermium* was announced in August 1955. Although Fermi died before the decision was made, Ghiorso himself wrote to Fermi's widow, Laura, to tell her the news. Einstein had died the previous April.

Prior to the final decisions about the names, several other suggestions had also been made. Several people thought element 100 should be called *centurium.* Naming one or both elements after national laboratories was considered. To recognize Los Alamos National Laboratory, names such as *losalium* and *alamosium* were proposed. To recognize Argonne National Laboratory, *argonnium* was proposed. Those, and similar, names, however, were not adopted.

Production of einsteinium and fermium takes place in nuclear reactors. Several years are required to accumulate measurable quantities. Probably less than a milligram of either element exists at any given time in the world. Today, there are 20 known isotopes of einsteinium, with half-lives ranging from eight seconds for Es-241 to 471 days for Es-252. Fermium has 21 known isotopes, with half-lives ranging from 0.8 millisecond for Fm-242 to 100 days for Fm-257. Fermium is the heaviest element that can be made using neutron capture methods. The method of beginning with targets even heavier than uranium—americium or curium, for example—does not yield heavier elements. Instead, target nuclei must be bombarded with heavier projectiles such as alpha particles.

Mendelevium

As elements become heavier, synthesizing them becomes increasingly more difficult. Beginning with element 101, synthesis takes place one atom at a time, often over a period of several months. The synthesis of 17 atoms of element 101 was announced in 1955 by Albert Ghiorso, Bernard G. Harvey, Gregory R. Choppin, Stanley G. Thompson, and Glenn Seaborg. Using the 60-inch cyclotron at Berkeley, an einsteinium 253 target on a backing of gold foil was bombarded with alpha particles as shown by the following reaction:

$$^{253}_{99}\text{Es} + ^{4}_{2}\text{He} \rightarrow ^{256}_{101}\text{Md} + ^{1}_{0}\text{n}.$$

Remarkably, the einsteinium target itself consisted of only one-billionth of a milligram of einsteinium. The discoverers chose to name element 101 *mendelevium* in honor of Dmitri Mendeleev, the Russian chemist who developed the periodic table of the elements in the form still used today.

Mendelevium was the last element that could actually be separated from any other elements present using ion exchange techniques. Subsequent elements have all been identified by the daughters produced by their radioactive decay. Mendelevium was also the last element that could be synthesized using projectiles as light as alpha particles since quantities of targets heavier than einsteinium were too insubstantial to be of any use.

Mendelevium has 21 known isotopes. The isotope with the shortest half-life is Md-245 at 0.9 millisecond. Mendelevium 258 has the longest half-life at 51.5 days.

Nobelium

After the discovery of mendelevium and the demonstration of successful "one atom at a time" techniques, scientists at various laboratories around the world began to search for element 102. In 1957, a research team at the Nobel Institute for Physics in Stockholm, Sweden, reported the discovery of element 102 and proposed the name *nobelium* in honor of Alfred Nobel (1833–96), the inventor of dynamite and the founder of the Nobel Prizes. The Swedish team's claim had to be withdrawn, however, when it was shown that they had misidentified the product of their reaction.

In October 1957, the heavy-ion linear accelerator (HILAC) at Berkeley was undergoing initial tests. By 1958, a team led by Albert Ghiorso, Torbjorn Sikkeland, John R. Walton, and Glenn Seaborg was able to produce element 102 by bombarding curium with carbon ions as shown in the following reaction:

$$^{246}_{96}\mathrm{Cm} + ^{12}_{6}\mathrm{C} \rightarrow ^{254}_{102}\mathrm{No} + 4\,^{1}_{0}\mathrm{n}.$$

The nobelium 254 atoms decayed into fermium 250 with a half-life of 49 seconds. Because the name *nobelium* had already been proposed by the Swedish scientists, the Berkeley discoverers chose to retain that name.

In 1966, teams of scientists led by Georgii Nikolaevich Flerov (1913–90) at the Dubna Laboratory in the Soviet Union were conducting parallel research. They reported the production of nobelium 256 using the following reaction:

$$^{238}_{92}\mathrm{U} + ^{22}_{10}\mathrm{Ne} \rightarrow ^{254}_{102}\mathrm{No} + 6\,^{1}_{0}\mathrm{n}.$$

In 1992, the IUPAC assessed the claims of discovery. The IUPAC concluded that the Berkeley team had made errors in its work, that the Dubna team had made the first conclusive discovery of element 102, and that Dubna deserved to be awarded priority of discovery. In 1994, the IUPAC officially adopted the name *nobelium.*

Twenty-one isotopes of nobelium are known. Nobelium 248 has the shortest half-life; it undergoes spontaneous fission with a half-life of less than one microsecond. Nobelium 259 is the longest-lived at 58 minutes.

Lawrencium

After the discovery of nobelium, there was room in the actinide series for one additional element—number 103. In 1961, using the HILAC at Berkeley, a team led by Albert Ghiorso, Torbjorn Sikkeland, Almon E.

("Bud") Larsh, and Robert M. Latimer bombarded a three-microgram californium target with boron ions—a mixture of boron 10 and boron 11. Using the reaction between californium 252 and boron 11 as an example, the production of element 103 occurred by the following reaction:

$$^{252}_{98}\text{Cf} + {}^{11}_{5}\text{B} \rightarrow {}^{258}_{103}\text{Lr} + 5\,{}^{1}_{0}\text{n}.$$

In 1965, the group at Dubna produced element 103 by bombarding an americium target with oxygen, as shown by the following reaction:

$$^{243}_{95}\text{Am} + {}^{18}_{8}\text{O} \rightarrow {}^{256}_{103}\text{Lr} + 5\,{}^{1}_{0}\text{n}.$$

Lawrencium 256 was identified by the following radioactive decay sequence:

$$^{256}_{103}\text{Lr} \rightarrow {}^{252}_{101}\text{Md} \rightarrow {}^{252}_{100}\text{Fm} \rightarrow {}^{248}_{98}\text{Cf}.$$

Once californium 248 had been reached, scientists had only to work backward to recognize lawrencium 256 as the starting material.

Element 103 was named lawrencium in honor of the Berkeley physicist Ernest Orlando Lawrence. *(New York Times Co./Getty Images)*

The Berkeley group proposed to name element 103 *lawrencium* in honor of Berkeley physicist Ernest Orlando Lawrence. This name was fitting since Lawrence was the inventor of the cyclotron, an achievement for which he received the Nobel Prize in physics in 1939. The IUPAC accepted the name and assigned lawrencium the symbol Lr.

Lawrencium has 14 isotopes. Lawrencium 252 has the shortest half-life at about 0.36 second. Lawrencium 262 is the longest-lived at 3.6 hours.

THE CHEMISTRY OF TRANSURANIUM ELEMENTS

The transuranium elements do not occur in nature and are radioactive in all forms. Progressing from left to right across the row in the table, half-lives became increasingly shorter, eventually becoming so short that only microgram quantities (or less) of the elements have ever been produced. The first compound of berkelium, for example, was $BkCl_3$; the sample weighed only three-billionths of a gram (three nanograms). Despite their minuscule quantities, the chemistry of the transuranium elements is surprisingly well documented.

A few general observations can be made. One such observation is that the first elements in the row exhibit chemistry resembling the chemistry of transition metals in that they form compounds and ions in multiple oxidation states.. The later elements in the row are more similar to their corresponding lanthanide elements in that the "+3" state is dominant. The transuranium elements are all active metals, which means they all oxidize rather easily. Their oxidation states exhibit a rough pattern. Oxidation states range from "+2" to "+7", with the "+3" state being the most common (similar to the lanthanides). The elements in the middle of the row exhibit a greater range of oxidation states than do the elements toward the beginning or the end of the row. The greater range in oxidation states of actinides compared to lanthanides demonstrates the difference between the "4f" electrons of the lanthanides versus the "5f" electrons of the actinides. The lanthanides' "4f" electrons are hidden more deeply inside the atom and are less available for chemical bonding. The "5f" electrons are more exposed and, thus, a greater number can participate in chemical bonding.

In the "+2" oxidation state, americium, californium, and berkelium form compounds with chlorine, bromine, and iodine. An example is americium chloride ($AmCl_2$). Einsteinium also forms "+2" halide compounds ($EsBr_2$ and EsI_2). Uncharacteristically of the heavier actinides, nobelium forms a "+2" fluoride and a "+2" oxide (NoF_2 and NoO, respectively).

The "+3" state tends to be the most common oxidation state. Most of the actinides in the "+3" state form compounds with fluorine, chlorine, bromine, and iodine. Plutonium forms a nitride (PuN). Americium, curium, einsteinium, fermium, mendelevium, and nobelium all form trioxides (Am_2O_3, Cu_2O_3, Es_2O_3, Fm_2O_3, Md_2O_3, and No_2O_3). Berkelium, californium, and einsteinium form oxychlorides (BkOCl, CfOCl, and EsOCl). Elements can also form complex ions in which the element is in the "+3" state. An example is the AmO_2^+ ion.

In the "+4" state, neptunium and plutonium form tetrafluorides (NpF_4 and PuF_4). Several elements, for example plutonium, neptunium, americium, curium, and berkelium, form dioxides (PuO_2, NpO_2, AmO_2, CmO_2, and BkO_2). Elements can be found as a "+4" ion, as occurs with americium (Am^{4+}), or as fluoride complexes, as also occurs with americium (AmF_8^{4-}).

Neptunium is the principal transuranium element exhibiting a "+5" oxidation state. This state is demonstrated in the pentoxide (Np_2O_5) and in an oxyfluoride ($NpOF_3$).

Several transuranium elements exhibit the "+6" state. For example, neptunium forms neptunium hexafluoride (NpF_6), and two "+6" oxyfluorides (NpO_2F_2 and $NpOF_4$). Berkelium and californium form trioxides (BkO_3 and CfO_3). Americium forms a positively charged ion in the "+6" state (AmO_2^{2+}).

Only neptunium and plutonium exhibit the "+7" state. The *oxyanions* NpO_5^{3-} and PuO_5^{3-} exist but are relatively unstable.

AMERICIUM IN THE HOME

One of the very few household devices known to be radioactive appears in nearly all homes, apartments, and businesses, and without it those dwellings would be much less safe to the occupants. Nearly all smoke detectors rely on the alpha particles produced by the decay of americium

Nearly all smoke detectors rely on the decay of americium 241. *(Danny E. Hooks/Shutterstock)*

241 bound in americium oxide (AmO_2) as a trigger to set off an alarm when smoke particles are present. When smoke is not present, these energetic alpha particles (E = 5.5 MeV) spend their time *ionizing* the nitrogen and oxygen molecules in the air, inducing an electric charge on the molecules. Within the detector a battery sets up a *potential difference* (voltage) between two plates. The charged ions travel toward the oppositely charged plates, setting up a current of ions that does not stop unless the air molecules cease to be ionized. That is precisely what happens when smoke particles intervene and absorb the alpha particles. The current stops, and an alarm is triggered.

The alpha particles, which are easily absorbed in the smoke detector's plastic container, and thereby stopped, are normally no hazard to humans or animals. Even if they should escape, they do not have sufficient energy to penetrate human skin. It could be hazardous, however, if someone were to breathe smoke from a burning detector. Another risk that should be considered is the eventual disposal of smoke detectors in landfills. The decay daughter product is neptunium 237, as shown below.

$$^{241}_{94}\text{Am} \xrightarrow{432\text{ y}} {}^{237}_{93}\text{Np} + {}^{4}_{2}\alpha + e^{-} + \gamma$$

While Am-241 is rather innocuous and has a fairly short half-life of 432 years, neptunium 237 is another story. In smoke detectors, it will make up about 5 percent in 30 years, which is not hazardous in households

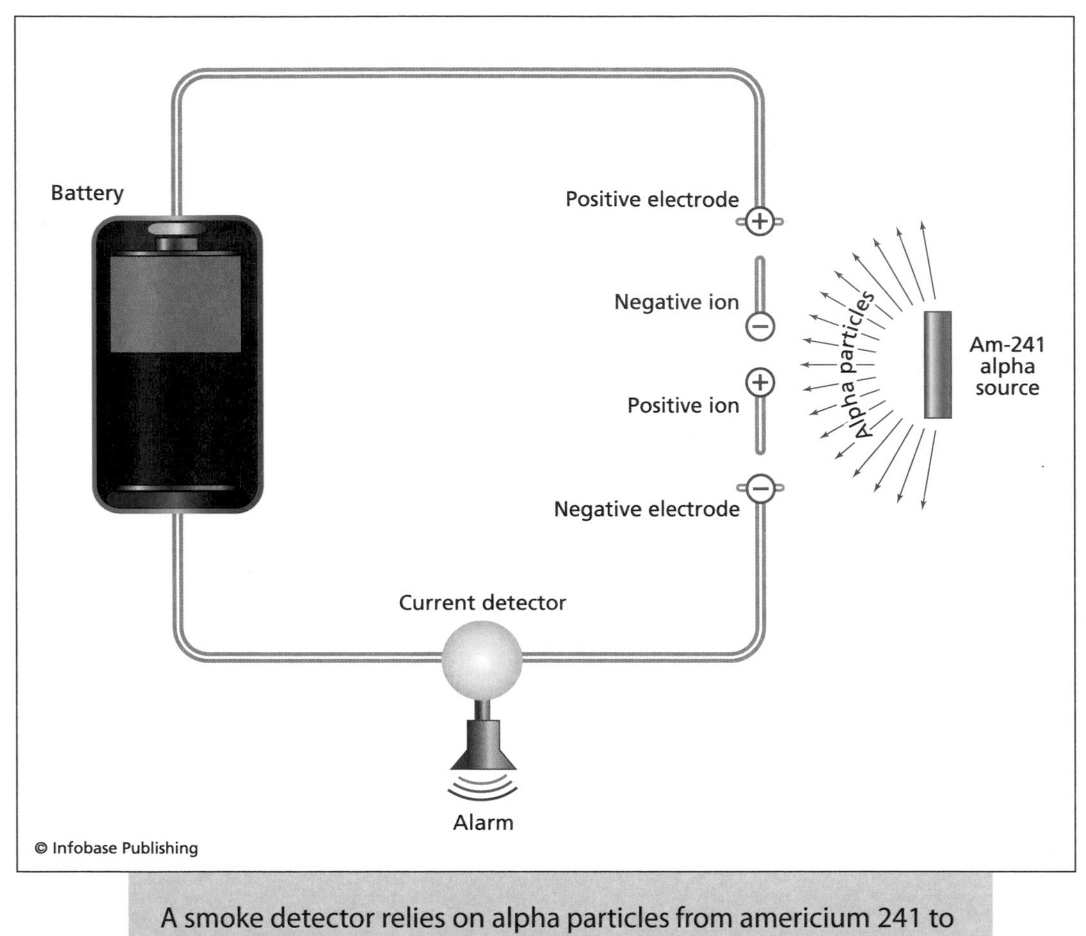

A smoke detector relies on alpha particles from americium 241 to trigger current flow.

because the lifetime of the electronics is usually shorter than that. However, Np-237 has a half-life of 2 billion years and many radioactive daughters (see decay chain on page 101). This means that smoke detectors will in the future pose a serious problem if relegated to unmonitored waste. Common sense indicates that recycling policies should be instituted in the near future for smoke detectors.

THERMONUCLEAR WEAPONS

For most of history, explosives have relied on chemical reactions. But in 1945 a new type of explosive changed the world in unforeseen and uncountable ways. Scientists working on the *Manhattan Project* in Los

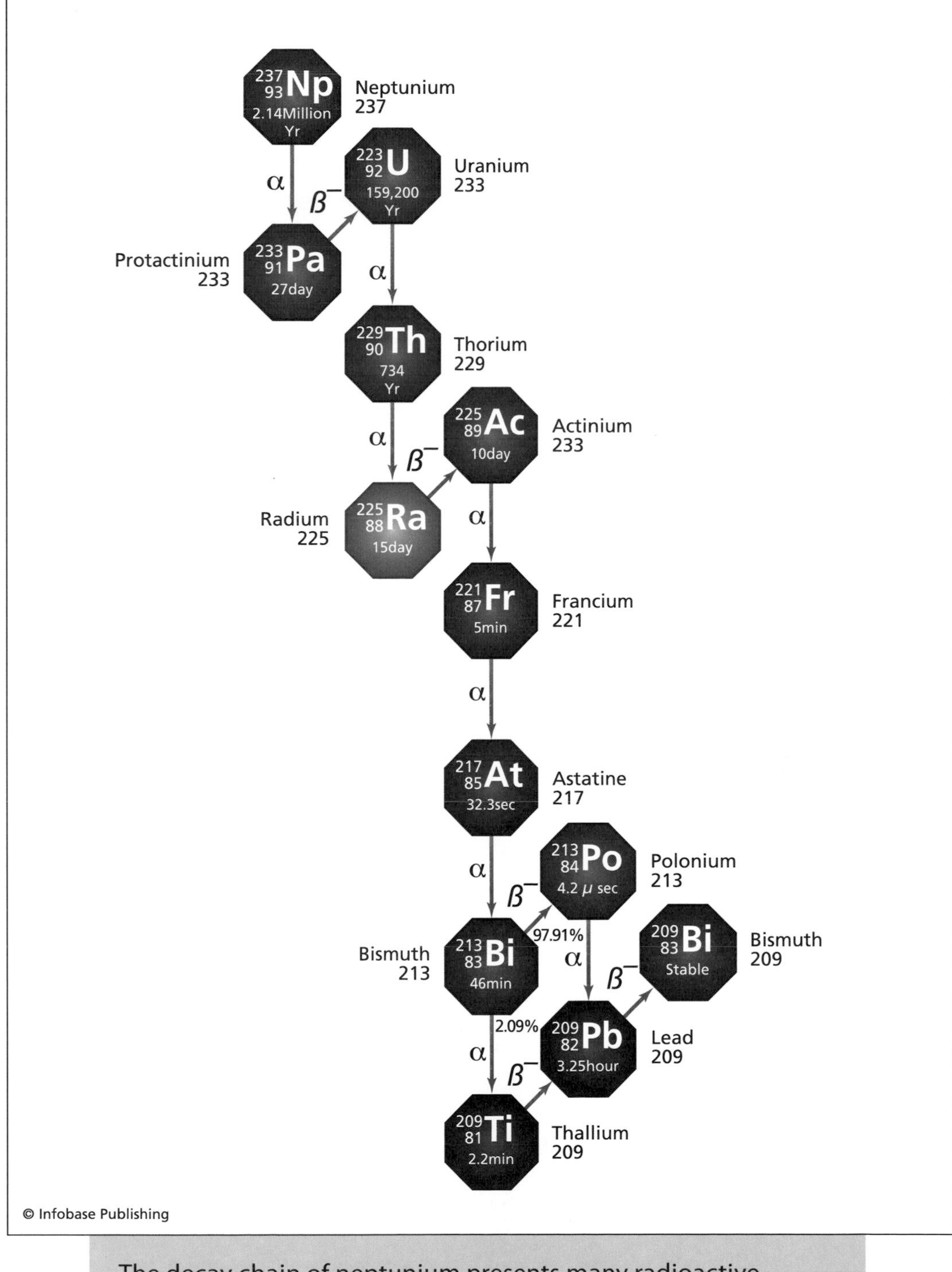

The decay chain of neptunium presents many radioactive daughters.

Alamos, New Mexico, figured out how to harness the power of nuclear fission to trigger a chain reaction in the limited geometry of a weapon that an aircraft could carry. This was an incredible feat of intellectual collaboration that led to stupendous explosions that destroyed two cities.

The bombs dropped on Hiroshima and Nagasaki, Japan, at the end of World War II, though devastating in the extreme, were orders of magnitude less destructive than what governments have available nowadays. Those early devices relied solely on the fission of U-235 and

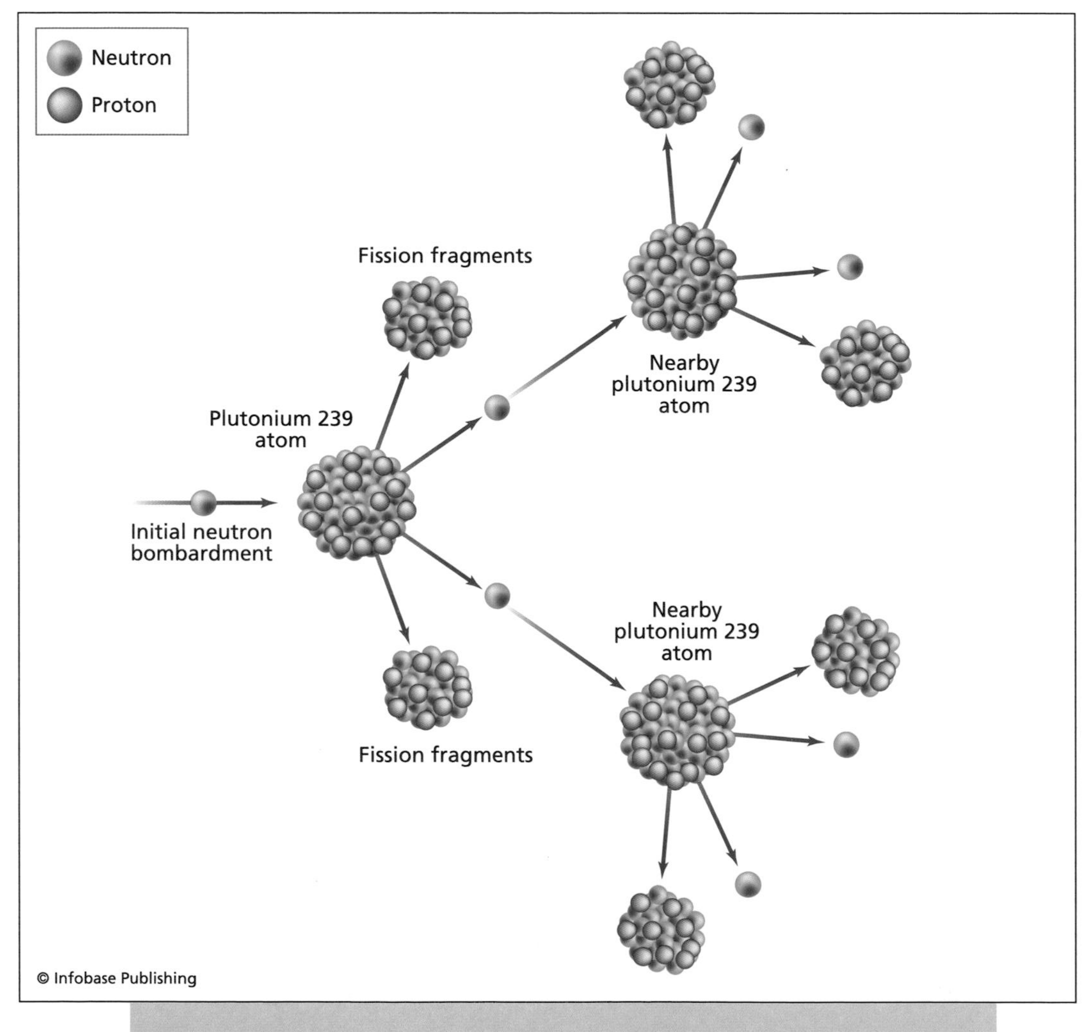

Each fission of Pu-239 produces two neutrons, each of which can then induce new fission reactions.

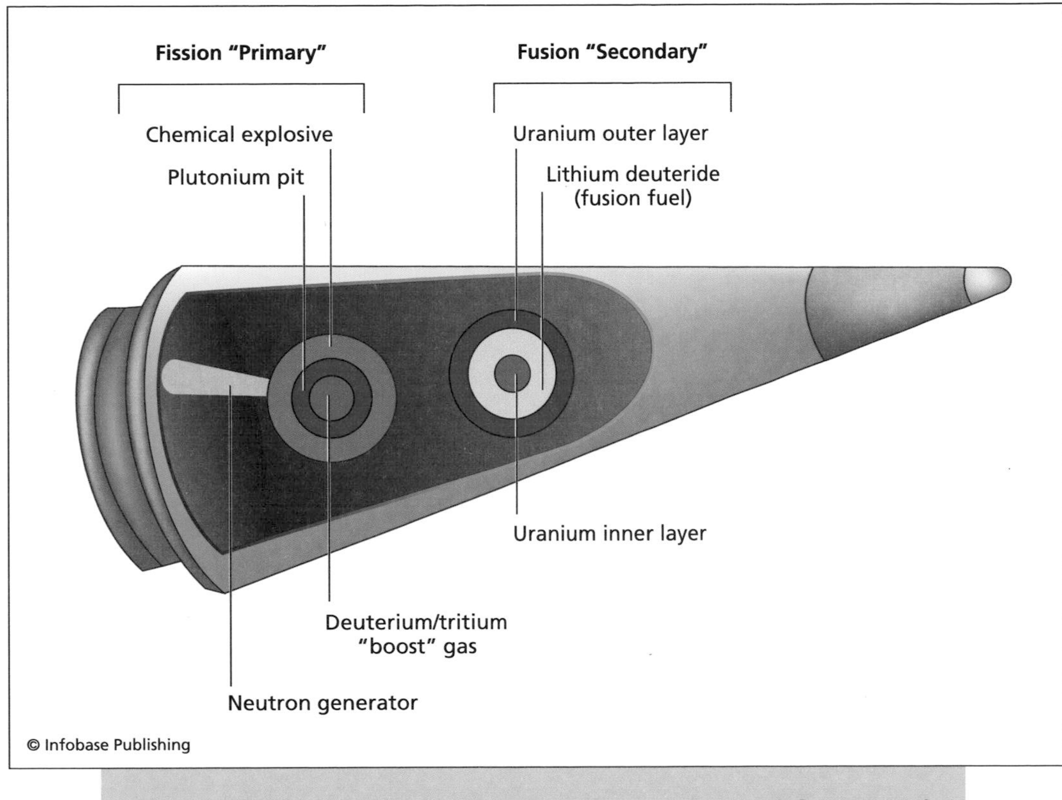

Modern thermonuclear weapons incorporate chemical, fission, and fusion reactions using hydrogen, lithium, uranium, and plutonium.

Pu-239, respectively. The power of the Hiroshima event was the equivalent of about 15 tons (13.6 metric tonnes) of TNT. For Nagasaki, it was about 20 tons (18.1 metric tonnes). How could such power come from such a small weapon? The answer lies in a chain reaction: Both U-235 and Pu-239 fission by interaction with slow neutrons, and—this is key—each reaction results in the emission of two neutrons. Anyone who understands the *exponential function* knows that *doubling time* is the most important aspect of growth. As shown in the figure on page 102, each of the two neutrons produces a fission that produces two more neutrons, and so on—a chain reaction leading to explosion within microseconds.

Modern thermonuclear weapons are 1,000 to 10,000 times more powerful than Little Boy or Fat Man (the nicknames given to the "atomic" bombs dropped on Japan). This spectacular increase in destructive potential was achieved by incorporating the fusion of hydrogen atoms, which is why these weapons are sometimes called H-bombs. To fuse hydrogen atoms requires extremely high pressure and temperature, such as exists in stellar cores. In a thermonuclear weapon, a primary fission reaction raises the temperature to the necessary 100 million Kelvin, at which point hydrogen atoms contained in *lithium deuteride* in the secondary region (see diagram, page 103) suddenly and explosively fuse. Additional uranium inner and outer layers in the secondary take advantage of neutrons produced by the fusion reaction, which then induce fission in the uranium for added explosive yield. This last step allows for lighter weapons that require less plutonium than the older versions.

Because of the overwhelming risks to world peace and the fate of humanity, world leaders are nearly unanimous in their agreement that

The Yucca Mountain area in Nevada was chosen by the U.S. Congress in 1987 as a nuclear waste repository, but environmental concerns have prevailed. The site will most likely never be used for that purpose. *(U.S. Department of Energy)*

such weapons should be completely abolished. In September 2009, the United Nations Security Council voted to work toward that goal. On April 8, 2010, the presidents of the United States and Russia signed an agreement that each country would reduce its stockpile to 1,550 deployed nuclear weapons—a 30 percent reduction over the previous strategic arms reduction treaty, which had expired in December 2009. Russia and the United States have been dismantling nuclear arsenals for at least a decade, but it is a time-consuming process, and dealing with the plutonium waste is difficult. While the uranium waste can be recycled into nuclear power plants, ^{239}Pu is a different story. First, high security is required when processing the plutonium to ensure that the material cannot be stolen for use in new weapons. Second, the waste is highly radioactive with a 24,000 year half-life, and no good method for retrieval and storage has yet been formulated. The problem of plutonium sequestration will remain an important area of research and development for decades into the future.

TECHNOLOGY AND CURRENT USES OF TRANSURANIUM ELEMENTS

While there are very few practical applications for most of the transuranium elements, many are used to some extent in research labs. Americium, for example, when combined with beryllium, serves as a reliable source for neutrons. As the americium decays, it gives off alpha particles that then interact with beryllium in the following reaction:

$$^{4}_{2}\alpha + ^{9}_{4}Be \rightarrow ^{1}_{0}n + ^{12}_{6}C.$$

The neutrons can also be used in nondestructive testing of equipment and in medical diagnostics, as neutrons are reflected and absorbed differently in different materials.

Plutonium is produced in significantly larger quantities than any of the other transuranium elements. Literally hundreds of tons of plutonium were made during the cold war, much of which is still stockpiled around the world. Plutonium is used as fissile material in nuclear reactors and in the primary stage of modern thermonuclear weapons, many of which are being dismantled according to a United Nations agreement for *nonproliferation*. The problem of sequestering plutonium from these weapons will remain an important area of research and development for decades into the future. Because of the potential that it could be diverted to terrorist activities, plutonium supplies are maintained

Many nuclear weapons are being dismantled according to a United Nations agreement for nonproliferation—a deactivated Minuteman II intercontinental ballistic missile. *(AP Photo/ National Park Service, HO)*

under strict security by the military and have limited civilian applications. Plutonium 238, for example, is used to make compact power generators for satellites.

The most familiar use of americium is its application as the detecting chemical in smoke detectors. Nearly all smoke detectors rely on the alpha particles produced by the decay of americium 241 bound in americium oxide (AmO_2) as a trigger to set off an alarm when smoke particles are present. The alpha particles, which are easily absorbed in the smoke detector's plastic container, and thereby stopped, are normally no hazard to humans or animals. Even if they should escape, they do not have enough energy even to penetrate human skin. It could be hazardous, however, if someone were to breathe smoke from a burning detector. Another risk which should be considered is the eventual disposal of smoke detectors in landfills.

Americium 241 can also be used as a diagnostic aid in analyzing bone disorders. In addition, americium is used as a source of neutrons and gamma rays. As a neutron source, americium has applications in oil well technology and in the determination of the density and moisture content of soils. As a gamma ray source, its importance is its use in portable equipment. A possible use that is still under investigation is to use the *metastable* form of americium 242 as an energy source for travel between Earth and other planets in the solar system.

Neptunium is made as a by-product of plutonium production and has no major commercial uses. Curium is used in pacemakers and in power supplies for satellites. Any of the *transcurium* elements are produced in such minuscule quantities that they have no uses other than in scientific research.

5

The Transactinide Elements

The transactinide elements are those elements with atomic numbers greater than 103. At lawrencium (element 103), the "5f" subshell becomes completely filled. Beginning with element 104, the "6d" subshell begins to fill. Because filling in a "d" subshell is characteristic of transition metals, the first 10 elements beginning with number 104 are rightly transition metals also. However, they are included in this book because, like the actinides, they are all artificial, radioactive elements, and the methods for their discoveries were derived from the methods for synthesizing the actinides.

The "6d" subshell is completely filled in upon reaching element 112. Beginning with element 113, the subshell that is being occupied by successively more electrons is the "7p," making those elements post-transition metals or nonmetals.

At the time of the writing of this book, the discovery of element 117 has not been definitively reported. Otherwise, there are 15 known transactinides. The first nine elements have been named—rutherfordium (Rf, element 104), dubnium (Db, 105), seaborgium (Sg, 106), bohrium (Bh, 107), hassium (Hs, 108), meitnerium (Mt, 109), darmstadtium (Ds, 110), roentgenium (Rg, 111), and copernicium (Cn, 112). The discoveries of elements 113–118 have been reported, but no names or symbols have yet been assigned. Most of these elements still require independent confirmation that the discoverers were indeed correct in their claims.

Because these elements have very short half-lives, properties such as boiling and melting point temperatures and electronic configuration are unknown, so tables of basic properties are not listed as they are for the other elements. The table of half-lives and decay modes on page 110 gives the information known about the basics of these elements.

In this chapter, the reader will learn about the discoveries of these elements and what little is known about their properties.

THE ORIGINS OF THE TRANSACTINIDES: HEAVY ION REACTIONS

While the transactinide elements are most likely synthesized in some very massive stars, astrophysicists have no way of observing this behavior because these elements are so short-lived, decaying to another element almost as soon as they are born. On Earth, the only way to synthesize the transactinides is to find a way to make elements with high mass stick together, which requires an exceptionally high-energy collision. Heavy atoms are difficult to accelerate, so usually a stationary target contains the heavier nuclei (such as uranium), whereas the projectile is an ionized version of a lighter atom (such as argon). Electromagnetic fields can be used to accelerate and deflect ionized atoms into the heavy target region. Identification of the new elements is quite challenging. The method involves the observation of decay products (alpha particles, for example) carrying specific energies that do not correspond to known elements. Expensive specialized detectors are required.

HALF-LIVES AND DECAY MODES OF THE LONGEST-LIVED ISOTOPES

ELEMENT	NUCLEON NUMBER	HALF-LIFE	DECAY MODE
104	255	1.4 seconds	SF, α
	257	4.8 seconds	α, SF
	259	3 seconds	α, SF
	261	78 seconds	α, SF
	263	10 minutes	α, SF
105	258	20 seconds	EC, α
	262	34 seconds	α, SF
	263	27 seconds	SF, α
	266	21 seconds	α
106	265	16 seconds	α
	266	21 seconds	α
107	264	0.44 second	α
	267	17 seconds	α
108	269	11 seconds	α
	270	3.6 seconds	α
	277	11.4 minutes	α
109	266	3.4 milliseconds	α
	268	70 milliseconds	α
110	280	7.4 seconds	SF
	281	1.1 minutes	α
111	272	1.5 milliseconds	α
112	283	3 minutes	SF
	285	10.7 minutes	α
113	283	~ 1 second	α
	284	~ 0.4 second	α
114	287	5 seconds	α
	289	~ 21 seconds	α
115	287	~ 46 milliseconds	α
	288	20–200 microseconds	α
116	289	47 milliseconds	α
117	293	14 milliseconds	α
	294	78 milliseconds	α
118	293	~1 millisecond	α

Key: α = alpha decay; β = beta decay; SF = spontaneous fission

DISCOVERY AND NAMING OF TRANSACTINIDES: EGO AND CONTROVERSY

After the synthesis of lawrencium, the heaviest actinide, theoreticians predicted that scientists should be able to synthesize even heavier elements. Just as there are *magic numbers* of electrons—2, 10, 18, 36, 54, etc.—that give atoms particular stability, there are magic numbers of protons and neutrons that give nuclei particular stability. For protons these numbers are 2, 8, 20, 28, 50, 82, 114, and 164. For neutrons they are 2, 8, 20, 28, 50, 82, 126, 184, and 196. Combinations of magic numbers of both protons and neutrons lead to nuclei that are especially stable against radioactive decay. For example, two protons and two neutrons give helium 4; 28 protons and 28 neutrons given iron 56, the most stable *nuclide* in the periodic table; and 82 protons and 126 neutrons give lead 208.

Recognizing that the next magic number of protons after lead is 114, it was reasonable to expect that element 114 might be more stable than the elements preceding it in the table. In other words, element 114 might exist as isotopes having significantly longer half-lives than the elements preceding it. In addition, other elements in the vicinity of 114 might share some of that stability.

The discoveries of the transactinide elements have resulted from work done at three principal laboratories: the Lawrence Berkeley National Laboratory at the University of California, Berkeley (LBNL); the Gesellschaft für Schwerionenforschung (GSI, Center for Heavy Ion Research) at Darmstadt, Germany; and the Joint Institute for Nuclear Research at Dubna, Russia (JINR). Because of competing claims of priority of discovery, naming these new elements has sometimes proven to be controversial. In all cases, the authority in naming rests with the International Union of Pure and Applied Chemistry, which consists of an international body of scientists that evaluates claims of discovery and evidence and makes final decisions.

Rutherfordium

Since the last element in the actinide group is lawrencium (number 103), scientists began searching for element 104 (eka-hafnium), which was expected to be a transition metal with properties analogous to zirconium and hafnium. In 1964, Georgiĭ Nikolaevich Flerov and coworkers at Dubna bombarded a target of plutonium with neon ions and claimed discovery of element 104 as shown by the following nuclear reaction:

$$^{242}_{94}Pu + ^{22}_{10}Ne \rightarrow ^{263}104 + ^{1}_{0}n.$$

The scientists in Dubna recommended the name kurchatovium (symbol Ku) after the late Russian nuclear physicist Igor Vasilevich Kurchatov (1903–60).

In 1969, a scientific team at Berkeley led by Albert Ghiorso was unable to reproduce the Soviets' results. Instead, they successfully produced atoms of element 104 by bombarding californium and curium targets with carbon and oxygen nuclei, as shown by the following nuclear reactions:

$$^{249}_{98}Cf + ^{12}_{6}C \rightarrow ^{257}104 + 4\,^{1}_{0}n;$$

$$^{249}_{98}Cf + ^{13}_{6}C \rightarrow ^{259}104 + 3\,^{1}_{0}n;$$

$$^{248}_{96}Cm + ^{16}_{8}O \rightarrow ^{258}104 + 6\,^{1}_{0}n.$$

Ghiorso's team recommended naming element 104 *rutherfordium* (symbol Rf) in honor of the New Zealand–born British physicist Ernest Rutherford (1871–1937). The IUPAC accorded the Berkeley scientists priority of discovery and adopted their recommendation of rutherfordium. Periodic tables published in the interval 1964–69, however, often listed element 104 as kurchatovium.

Rutherfordium belongs to the titanium group of elements and is located under hafnium in the periodic table. Sufficient chemistry has been done with rutherfordium to confirm that it is *tetravalent* (it exhibits chemistry in the "+4" oxidation state) and therefore more closely resembles zirconium and hafnium than it does the actinides. Fourteen isotopes have been synthesized; none of them have very long half-lives. The half-life of rutherfordium 254 is only 23 microseconds. Rutherfordium 265 is the longest lived at 13 hours.

Dubnium

Like element 104, element 105's discovery and naming also involved controversy. In 1968, Georgiĭ Nikolaevich Flerov and his coworkers at Dubna reported the synthesis of two isotopes of element 105 (eka-tantalum) using the following reactions between americium and neon ions:

$$^{243}_{95}\text{Am} + {}^{22}_{10}\text{Ne} \rightarrow {}^{260}105 + 5\,{}^{1}_{0}\text{n};$$

$$^{243}_{95}\text{Am} + {}^{22}_{10}\text{Ne} \rightarrow {}^{261}105 + 4\,{}^{1}_{0}\text{n}.$$

However, when their data were analyzed by other groups, the conclusion was that the Soviet group was in error.

In 1970, Albert Ghiorso and his coworkers Matti Nurmia, Jim Harris, Kari Eskola, and Pirkko Eskola at Berkeley reported the synthesis of element 105 by bombarding a californium 249 target with nitrogen 15 ions, as shown in the following equation:

$$^{249}_{98}\text{Cf} + {}^{15}_{7}\text{N} \rightarrow {}^{260}105 + 4\,{}^{1}_{0}\text{n};$$

The group produced two more isotopes of element 105 the following year. The results of the Berkeley scientists were accepted, and they were credited with priority of discovery.

The Soviet group recommended the name *nielsbohrium* in honor of the Danish physicist Niels Bohr (1885–1962). The American group recommended the name *hahnium* (Ha) in honor of the German chemist Otto Hahn. At first, the name hahnium and symbol (Ha) were accepted, and those were the name and symbol used in the scientific literature for almost 25 years. In the mid-1990s, however, the IUPAC decided to name element 105 *dubnium* after the town in Russia where the JINR is located. The symbol the IUPAC adopted was Db. At the same time, the IUPAC shortened the name *nielsbohrium* to *bohrium* and assigned that name to element 107 (symbol Bh).

Experiments performed at Berkeley in the 1980s by Kenneth E. Gregorich, Darleane Hoffman, and coworkers demonstrated that dubnium is *pentavalent.* In other words, it belongs to the vanadium family of elements and behaves similarly to niobium and tantalum in forming compounds in the "+5" oxidation state. Remarkably, experiments in aqueous solution could be done despite the fact that dubnium's longest-lived isotope is Db-262, with a half-life of only 34 seconds.

Seaborgium

Element 106 (eka-tungsten) is the heaviest known member of the chromium group. The synthesis of element 106 was first reported in 1974 at the Lawrence Berkeley Laboratory by a team lead by Albert Ghiorso, Kenneth Hulet, and Glenn Seaborg. A target of californium 249 was bombarded with oxygen 18 as shown in the following reaction:

$$^{249}_{98}\mathrm{Cf} + ^{18}_{8}\mathrm{O} \rightarrow ^{263}106 + 4\,^{1}_{0}\mathrm{n}.$$

A reasonable question to ask is: How are scientists able to identify elements that are made only one atom at a time and that exist so briefly before undergoing radioactive decay? The answer is that a radioactive decay series is observed. Suppose one starts with Sg-263. It undergoes an alpha decay to form Rf-259. Rutherfordium decays to No-255. These decays are summarized in the following series:

$$^{263}_{106}\mathrm{Sg} \xrightarrow{\alpha} ^{259}_{104}\mathrm{Rf} \xrightarrow{\alpha} ^{255}_{102}\mathrm{No}.$$

Each alpha decay occurs with a previously known half-life. In addition, each alpha particle has a characteristic energy. As soon as scientists begin detecting alpha decays that they can attribute to known isotopes, they can work backward in the series to identify what the initial isotope must have been. The genetic relationship between isotope 263 of element 106 that was produced, its daughter $^{259}_{104}\mathrm{Rf}$, and its granddaughter $^{255}_{102}\mathrm{No}$ is considered positive proof by nuclear scientists that element 106 had in fact been synthesized.

It was several years before the discovery was accepted, but in 1993 a Berkeley team led by Ken Gregorich and Darleane Hoffman confirmed the results of the original team. On the 20th anniversary of the element's discovery, the Berkeley scientists were given permission by the IUPAC to propose a name for the new element. Because Berkeley chemist and Nobel laureate Glenn T. Seaborg (1912–99) had been the discoverer or codiscoverer of most of the transuranium elements beginning with plutonium in 1940, the discoverers of element 106 met without Seaborg's knowledge and decided to honor him by naming element

106 *seaborgium* (symbol Sg). At the March 1994 national meeting of the American Chemical Society, Ken Hulet announced before the Division of Nuclear Chemistry that the name *seaborgium* for element 106 had been proposed to the IUPAC.

Although it was not completely unprecedented for an element to be named for a living person (Einstein and Fermi were both still alive when it was proposed that elements 99 and 100 be named for them), the IUPAC initially rejected the proposal on the grounds that Seaborg was still living. There was so much support for the name, however, that the IUPAC finally agreed to it. Seaborg lived to see the name *seaborgium* become official before his death in 1999 due to complications from a stroke.

Seaborgium isotopes with longer half-lives than $^{263}_{106}Sg$ have been synthesized. Seaborgium 265 has a half-life of about 15 seconds, and seaborgium 271 has a half-life of about two minutes. Those half-lives are sufficient to allow seaborgium's chemistry to be studied. Seaborgium exhibits chemistry similar to the "+6" oxidation-state chemistry of its *homologues* chromium, molybdenum, and tungsten, forming, for example, the compound SgO_2Cl_2 in the gas phase (which is similar to CrO_2Cl_2, MoO_2Cl_2, and WO_2Cl_2).

Bohrium

The discovery of element 107 (eka-rhenium) was reported in 1981 by a team of scientists led by Peter Armbruster (1931–) and Gottfried Münzenberg (1940–) at GSI in Darmstadt, Germany. A different approach was taken from the usual bombardment of *transfermium* elements with lightweight projectiles. Instead, scientists began using lead and bismuth targets, which were bombarded with medium-weight projectiles. In this case, the target was bismuth 209, and the projectile was chromium 54, as shown in the following reaction:

$$^{209}_{83}Bi + ^{54}_{24}Cr \rightarrow {}^{262}107 + ^{1}_{0}n.$$

Five atoms of $^{262}107$ were made, all of which underwent alpha decays to form lighter isotopes, as shown by the following example:

$$^{262}107 \rightarrow ^{258}_{105}Db + ^{4}_{2}He \rightarrow ^{254}_{103}Lr + ^{4}_{2}He.$$

Bohrium is named in honor of the Danish physicist Niels Bohr. *(AP Photo/Alan Richard)*

Again, the genetic relationship between isotope 262 and its daughter and granddaughter confirmed the production of element 107.

Element 107 was originally named *nielsbohrium* in honor of the Danish physicist Niels Bohr, but in 1997 the IUPAC shortened the name to *bohrium* (symbol Bh). Element 107 is the heaviest member of Group VIIA. Only a few atoms with very short half-lives have ever been produced (Bh-272 has the longest half-life, about one second), so that nothing is known at the present time about bohrium's chemistry.

Hassium, Meitnerium, and Darmstadtium

All three of these elements were produced artificially at GSI in Darmstadt, Germany. Hassium (symbol Hs) is element 108 (eka-osmium). It was first made in 1984 by Peter Armbruster, Gottfried Münzenber, and their coworkers by bombarding bismuth 209 with manganese 55, and lead 207 and lead 208 with iron 58, as shown:

$$^{209}_{83}Bi + ^{55}_{25}Mn \rightarrow ^{263}108 + ^{1}_{0}n$$

$$^{207}_{82}Pb + ^{58}_{26}Fe \rightarrow ^{264}108 + ^{1}_{0}n$$

$$^{208}_{82}Pb + ^{58}_{26}Fe \rightarrow ^{265}108 + ^{1}_{0}n$$

No observable quantity of these isotopes can be produced because they are alpha-emitters with half-lives of two milliseconds or less. Subsequently, a number of other isotopes have been produced. Hassium 277 has the longest half-life observed so far—11 minutes.

The name *hassium* was chosen and was derived from the Latin word Hassia, which refers to the German state of Hesse.

Meitnerium (symbol Mt) is element 109 (eka-iridium). It was discovered in 1982 by the same team at GSI when one atom of element 109 was produced by bombarding a bismuth target with a beam of iron nuclei, as shown by the following nuclear reaction:

$$^{209}_{83}Bi + ^{58}_{26}Fe \rightarrow ^{266}109 + ^{1}_{0}n.$$

Atoms of several more isotopes of meitnerium have been synthesized, but they do not last long. The longest half-life observed to date is only 0.72 second (Mt-276).

The name *meitnerium* was chosen to honor the Austrian nuclear physicist Lise Meitner (1878–1968), who made a significant contribution to the discovery of nuclear fission.

Element 110 (eka-platinum) was discovered in 1994 at GSI by Armbruster, Münzenber, Sigurd Hofmann, and coworkers. Element 110 was produced by bombarding a lead target with beams of nickel nuclei, which included both nickel 62 and nickel 64. Atoms of darmstadtium were produced by the following nuclear reactions:

$$^{208}_{82}Pb + ^{62}_{28}Ni \rightarrow ^{269}110 + ^{1}_{0}n$$

$$^{208}_{82}Pb + ^{64}_{28}Ni \rightarrow ^{271}110 + ^{1}_{0}n$$

Several additional isotopes of darmstadtium have been discovered. Most half-lives are less than one millisecond. Ds-281 has the longest

Meitnerium is named in honor of the Austrian nuclear physicist Lise Meitner. *(AP Photo/MPG HO)*

half-life—11 seconds. Darmstadtium (symbol Ds) is named after the city in which it was discovered.

Darmstadtium 269 undergoes alpha decay to form Hs-265. Hassium, in turn, undergoes another alpha decay to form Sg-261, which in turn decays to Rf-247. Rutherfordium decays to No-253. These decays are summarized in the following series:

$$^{269}_{110}\text{Ds} \xrightarrow{\alpha} {}^{265}_{108}\text{Hs} \xrightarrow{\alpha} {}^{261}_{106}\text{Sg} \xrightarrow{\alpha} {}^{257}_{104}\text{Rf} \xrightarrow{\alpha} {}^{253}_{102}\text{No}.$$

Each alpha decay occurs with a previously known half-life. In addition, each alpha particle has a characteristic energy. As soon as scientists begin detecting alpha decays that they can attribute to known isotopes, they can work backward in the series to identify what the initial isotope must have been.

Roentgenium

Element 111 (eka-gold) was also discovered at GSI in 1994 by the same group of scientists who discovered darmstadtium by bombarding a bismuth 209 target with nickel 64 ions, as shown:

A *Life* magazine cartoon from 1896 lampoons the discovery of the X-ray by the scientist Wilhelm Konrad Roentgen, for whom roentgenium is named. *(AP Photo)*

$$^{209}_{83}\text{Bi} + ^{64}_{28}\text{Ni} \rightarrow ^{272}111 + ^{1}_{0}\text{n}.$$

Roentgenium (symbol Rg) is named after Wilhelm Conrad Roentgen (1845–1923), the German physicist who discovered X-rays in 1895. Roentgenium 272 has a half-life of about two milliseconds. The longest-lived isotope is Rg-280 with a half-life of about 3.6 seconds.

Elements 112–118

The search for element 112 (eka-mercury) was conducted over a 24-day period in January and February 1996, with the creation of the first atom of 112 occurring on February 9. A lead 208 target was bombarded with zinc 70 nuclei, resulting in a product with a mass number of 277 as shown in the following equation:

$$^{208}_{82}\text{Pb} + ^{70}_{30}\text{Zn} \rightarrow ^{277}112 + ^{1}_{0}\text{n}.$$

Atoms of element 277 are alpha-emitters with a half-life of only 0.7 millisecond, so the first atom was very short-lived.

In 2000, the team at GSI repeated the experiment and produced another atom of Uub-277. The experiment was duplicated at the Japanese laboratory RIKEN (Rikagaku Kenkyūsho, which translates as the Institute of Physical and Chemical Research). In 2001, and again in 2003, the IUPAC decided that GSI's evidence was insufficient to support their claim of discovery. In May 2009, however, scientists at both GSI and RIKEN presented more confirmation and the IUPAC accepted the claim and recognized GSI as having priority of discovery.

As of 2009, five isotopes of element 112 have been made. The longest-lived isotope has a mass number of 285 with a half-life of 29 seconds. In 2010, the IUPAC officially decided to name element 112 after Nicolaus Copernicus, the famous 16th-century Polish astronomer whose theory of a heliocentric universe revolutionized science. Element 112 is now *copernicium* with the symbol Cn.

In February 2004, a team of scientists from LLNL in California and JINR in Dubna, Russia, reported the discoveries of two new elements—numbers 113 (eka-thallium) and 115 (eka-bismuth). The bombardment of an americium 243 target with calcium 48 nuclei resulted in the production of four atoms of element 115, as shown in the following reaction:

$$^{243}_{95}\text{Am} + ^{48}_{20}\text{Ca} \rightarrow ^{288}115 + 3\ ^{1}_{0}\text{n}.$$

The atoms that were made had a mass number of 288 and lasted about 0.1 second before decaying into element 113, as shown:

$$^{288}115 \rightarrow ^{284}113 + ^{4}_{2}\text{He}.$$

The atoms of $^{284}113$ themselves decayed with a half-life of about one-half of a second.

Four isotopes of element 113 are known. All of them are alpha-emitters with half-lives less than one second.

Although the experimental data appears convincing, the data are waiting for independent confirmation. Until that time, no permanent names will be assigned, but during the interim period scientists will use the temporary names *ununtrium* (Uut) for element 113 and *ununpentium* (Uup) for 115.

The first claim for the discovery of element 114 came in early 1999 from scientists at Dubna, Russia, who reported the synthesis of a single atom. Subsequently, the syntheses of a few more atoms were reported. (Each experiment ran for just more than a month.) Element 114 is made by bombarding a plutonium 244 target with calcium 48 nuclei, as shown in the following reactions:

$$^{244}_{94}\mathrm{Pu} + ^{48}_{20}\mathrm{Ca} \rightarrow {}^{288}114 + 4\,^{1}_{0}\mathrm{n}$$

$$^{244}_{94}\mathrm{Pu} + ^{48}_{20}\mathrm{Ca} \rightarrow {}^{289}114 + 3\,^{1}_{0}\mathrm{n}.$$

Both isotopes decay with half-lives of less than one minute: five seconds for mass number 288 and 30 seconds for mass number 289. Although in one sense these half-lives are short, they are actually quite long for transactinide elements, which often have half-lives of milliseconds or less. For several years, nuclear chemists had been predicting that heavy elements might be found with half-lives of seconds or more, so the observation of these two half-lives generated considerable excitement.

Later, element 114 was also observed following the synthesis of element 118, which then very rapidly undergoes a series of alpha decays, as shown in the following sequence:

$$^{293}118 \rightarrow {}^{289}116 \rightarrow {}^{285}114 \rightarrow {}^{281}112.$$

The half-life of each decay is less than one millisecond. Since only one atom of element 118 is produced at a time, for the near future at least, it is unlikely that any weighable quantities of these elements can be formed.

It was subsequently concluded that the Dubna scientists' initial 1999 report was in error. However, they had made other observations in

2000 to 2004 that held up under scrutiny. In 2009, scientists at the Lawrence Berkeley National Laboratory in Berkeley, California, confirmed the later reports that came from Dubna. The Berkeley scientists made two atoms of element 114 using the Dubna scientists' later methods, thus confirming Dubna's priority of discovery. The temporary name for element 114 is ununquadium (Uuq).

Elements 116 and 118, temporarily named ununhexium (Uuh) and unuoctium (Uuo) respectively, have a checkered history. In 1999, scientists at the Lawrence Berkeley National Laboratory reported the discoveries of elements 116 and 118. However, in 2000 and 2001, when the experiments were repeated at Berkeley and simultaneously at the Institute for Heavy Ion Research at Darmstadt, Germany, and the Atomic Energy Research Institute in Japan, scientists were unable to confirm the same results. Puzzled, the Berkeley scientists re-examined the 1999 data and reported that one of the scientists on the earlier team allegedly had fabricated the data that suggested that elements 116 and 118 had been synthesized. Consequently, the Berkeley scientists retracted the claim.

As regrettable as this incident may have been, it does serve as a good example of how scientific research works. Any discovery is subject to peer review and replication by other scientists. Only when an experiment has been replicated and the results accepted by others in the scientific community is the discovery considered to be valid. In the case of discovering new elements, any element must be synthesized by more than one team of scientists before the international community will officially include the new element in the periodic table.

Subsequently, in 2002 and again in 2005, elements 116 and 118 were produced by a collaborative effort of scientists at the Joint Institute for Nuclear Research at Dubna and the Lawrence Livermore Laboratory. The procedure they used was to bombard a californium 249 target with calcium 48 nuclei, as shown in the following equation:

$$^{249}_{98}\text{Cf} + ^{48}_{20}\text{Ca} \rightarrow {}^{294}118 + 3\,^{1}_{0}\text{n}.$$

The Uuo-294 isotope quickly undergoes alpha decay to form Uuh-116, as shown in the following equation:

$$^{294}118 \rightarrow {}^{290}116 + ^{4}_{2}\text{He}.$$

The half-life for the decay is only 1.29 milliseconds. Since element 118 is made only one atom at a time, and element 116's half-life is only 14.4 milliseconds, neither element has a chance to accumulate.

In addition to the formation of element 116 as the daughter of the alpha decay of element 118, element 116 was also produced in 2000 by the Russian and American collaborators who made element 118. Their method for making element 116 was to bombard a curium 248 target with calcium 48 ions, as shown in the following equation:

$$^{248}_{96}Cm + ^{48}_{20}Ca \rightarrow ^{292}116 + 4\,^{1}_{0}n.$$

The half-life of Uuh-292 is only 47 milliseconds.

In April 2010, a collaboration of scientists from Russia and the United States reported that they had been able to produce six atoms of element 117 (temporarily named ununseptium, Uus) at JINR in Dubna, Russia. This elusive element was finally manufactured in the laboratory by bombarding a target made of berkelium 249 atoms—provided by Oak Ridge National Laboratory—with calcium 48 ions, resulting in the following reactions:

$$^{48}_{20}Ca + ^{249}_{97}Bk \xrightarrow{78ms} ^{294}117 + 3^{1}_{0}n$$

$$^{48}_{20}Ca + ^{249}_{97}Bk \xrightarrow{14ms} ^{293}117 + 4^{1}_{0}n$$

Measured decay times appear above the reaction arrows. Five atoms of Uus-293 were detected and one of Uus-294. As is always the case, official naming will not take place until after the discovery is formally reproduced. The JINR scientists will most likely continue working to synthesize elements 119 and 120 in the future. If element 120 is discovered, it should be an alkaline earth element, with properties similar to the properties of barium and radium.

SHORT LIVES, FEW USES OF THE TRANSACTINIDES

Only minute quantities have been produced of the elements in the bottom row of the transition metals. Because of their isotopes' very short half-lives, it is unlikely that macroscopic quantities will ever be made. Nonetheless, nuclear chemists working at the Berkeley and Livermore laboratories in California, at GSI in Germany, at Dubna in Russia, and at RIKEN in Japan, continue to produce samples of these elements, even if only at the rate of one atom at a time. A single atom is sufficient, however, to permit the study of its properties. Even with projects running 24 hours a day, seven days a week, often for weeks or months at a time, chemists may only be able to produce one atom every two weeks or so—and it decays within seconds. That is enough, however, to permit studies of elements such as rutherfordium (element 104) and dubnium (element 106). While results of these experiments are inconclusive, evidence suggests that these *transactinide* elements do exhibit properties similar to the elements lying above them in the periodic table. Further experiments may be expected to continue.

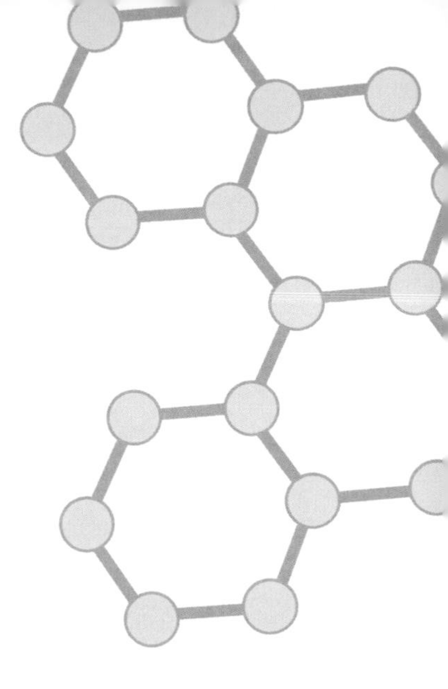

6

Future Directions

It is important for scientists to think ahead, to attempt to guess what areas are ripe for investigation and which may be bound for oblivion. If one considers recent remarkable leaps in information science, medicine, and particle physics that have materialized in the past century, it is clear that predictions are bound to be less impressive than eventualities, but there are some obvious starting points. Some of those, especially related to the lanthanides and actinides, are suggested here.

NEW PHYSICS

The near future of the lanthanides and actinides will most likely center on considerations regarding the environment and the promotion of cleaner energy. In particular, the mining of rare earth metals for use in wind turbines and hybrid vehicles will be important, as will the development of thorium reactors.

In order to develop energy sources that do not emit carbon dioxide, materials with good temperature stability and magnetic flux retention are needed. Neodymium and samarium are the best elements to supply these properties when alloyed with other metals. Permanent magnets that incorporate neodymium or samarium are being used in wind turbines and in hybrid and electric vehicles. While rare earth alloys were invented in the United States, production of these materials has in the last decade been outsourced almost exclusively to China, which currently provides nearly 100 percent of the world's rare earth elements. Demand worldwide is steadily increasing, while Chinese exports of rare earth oxides are diminishing, owing to its focus on internal development of alternative energy sources. Estimates of global energy needs indicate that annual neodymium production will need at least to double in order to keep pace with demand.

The global energy industry will also rely on more nuclear power, which is at an important juncture. The problem of waste product disposal has not been solved, whereas nuclear reactors emit no greenhouse gases. Investigations continue into the feasibility and efficiency of modifying enriched uranium reactors to enable thorium fission power. Thorium is widely available in many countries in deposits that require less environmentally destructive mining techniques than uranium ore. Radioactive waste by-products from thorium reactors would have short half-lives (on the order of decades), so safe storage is not such a problem, and bomb production from the waste products would be exceedingly more difficult and expensive than such production from plutonium. The switch to thorium will not be simple, however. Companies will need an incentive to invest in mining. The push will most likely need to come from the public sector as people learn of this option.

The outlook on uranium into the distant future will involve waste cleanup and sequestration. Mountains of tailings from decommissioned uranium mines continue to threaten the environment. Problems with mining and reactor waste abound, while legislation moves at a pace that does not keep up with the needs of human health or the environment. The controversy over the use of depleted uranium in weapons and armor will continue until governments are able to sort out facts from rumors regarding the health risks and legal precedents.

Dating stone tools millions of years old (such as the one pictured) is sometimes aided by the alignment of magnetic grains in surrounding rock. A uranium reactor at Earth's core might simplify the explanation of why the planet's magnetic field changes over the course of millennia. *(AP Photo/ The State University of New Jersey Rutgers)*

While some uranium waste can be recycled into nuclear power plants, ^{239}Pu is a different story. High security is required when processing the plutonium to ensure that the material cannot be stolen for use in new weapons. Additionally, the waste is highly radioactive with a 24,000-year half-life, and no good method for retrieval and storage has yet been formulated. The problem of plutonium sequestration will remain an important area of research and development for decades into the future.

An intriguing hypothesis currently being examined is the possibility that Earth's core may contain, not molten iron—the standing theory—but a molten uranium natural reactor. While far from confirmed, this condition could explain in a fairly simple way the shifting of Earth's magnetic field in a seemingly random fashion. Much more study will be needed to test this unusual premise.

Regarding the science behind the formation of lanthanides and actinides in stars, much remains to be discovered. In particular, continuing research on the excesses of such extremely heavy elements in chemically peculiar stars will advance knowledge about the thermodynamics and mass-dependence of nucleosynthesis.

NEW CHEMISTRY

Chemists often discuss the notion of whether or not the periodic table will ever be considered "finished." The answer to that question is unknown but has been the subject of scientific investigation for more than 70 years. As of 2010, nuclear scientists continue to work along several lines of inquiry: to confirm reported claims for new elements; to produce heavier isotopes of elements that have already been reported, and that should prove to have longer half-lives; and to synthesize new elements beyond element 118.

Current efforts are aimed at the synthesis of element 120. Element 120 (and 119, if also discovered) would begin the 8th period of the periodic table. Element 121 would be a new transition metal (eka-actinium) with a valence electron in the "7d" subshell. Elements 122 to 153 would be *superactinides,* comprising a new row below the actinides. The first 14 superactinides would be the first elements with outermost electrons in "g" orbitals (the "5g" subshell), while the next 18 superactinides would have outermost electrons in the "6f" subshell. Element 154 (eka-rutherfordium) would be a transition metal following element 121. The remainder of the 8th period would be elements 155 to 168 and would fill in the "7d" and "8p" subshells. Element 168 would be a noble liquid since it would be in the same family as the *noble gases,* but would be expected to be a liquid at 1 atm pressure and room temperature (77°F [25°C]).

It simply is unknown at this time whether or not it will ever be possible to produce elements that are that heavy. Extremely high energies would be required to produce such heavy elements, requiring expensive technologies that presently do not exist. Any prospect of ever doing so may lie years into the future. A ray of hope exists, however. As early as the 1940s, it was shown that the collection of neutrons and protons in atomic nuclei are particularly stable when the numbers of neutrons or protons (the so-called magic numbers) through uranium are 2, 8, 20, 28, 50, and 82, with larger numbers expected for transuranium elements. Doubly-magic nuclei are ones in which both the numbers of neutrons and protons are magic numbers, conferring on those isotopes extra stability against radioactive decay.

When isotopes are graphed with proton numbers on one axis and neutron numbers on the other axis, a "peninsula" of stable isotopes appears to form surrounded on both sides by a "sea" of unstable isotopes. Glenn Seaborg visualized the peninsula as pointing toward *islands of stability* amid the *sea of instability.* These "islands" would consist of elements with doubly magic numbers of protons and neutrons; these elements are expected to have half-lives (possibly as long as millions of years) that would be considerably longer than the millisecond half-lives of lighter artificial elements. Element 114 was expected to be such an island, although it proved not to be as stable as was predicted. There are competing, and conflicting, predictions about where islands of stability should occur. Various theories point to elements 120, 124, 126, 168, and 184.

In the meantime, the discovery of element 112 (eka-mercury) by Sigurd Hofmann and coworkers at Darmstadt has been confirmed and element 112 given a tentative name—*copernicium,* in honor of the 16th-century Polish astronomer Nicolaus Copernicus (1473–1543), who proposed that Earth revolves around the Sun, and not vice versa. Although only a few atoms of copernicium have been made so far, early experiments suggest that its properties are in fact similar to the properties of mercury, and that it, too, may be a liquid at room temperature.

The uncertainty of what scientists will someday find is what makes the quest for scientific knowledge so exciting. What seems highly improbable or even impossible today could become commonplace tomorrow. The fundamental nature of matter underlies the arrangement of elements in the periodic table and makes the table so useful. Extending the periodic table to superheavy elements and matching their actual properties with their predicted properties can only reinforce scientists' understanding of the matter that is the "stuff" of planets and stars and galaxies—and life itself.

SI Units and Conversions

UNIT	QUANTITY	SYMBOL	CONVERSION
Base units			
meter	length	m	1 m = 3.2808 feet
kilogram	mass	kg	1 kg = 2.205 pounds
second	time	s	
ampere	electric current	A	
kelvin	thermodynamic temperature	K	1 K = 1°C = 1.8°F
candela	luminous intensity		
mole	amount of substance	mol	
Supplementary units			
radian	plane angle	rad	(pi / 2) rad = 90°
steradian	solid angle	sr	
Derived units			
coulomb	quantity of electricity	C	
cubic meter	volume	m^3	1 m^3 = 1.308 $yards^3$
farad	capacitance	F	
henry	inductance	H	
hertz	frequency	Hz	
joule	energy	J	1 J = 0.2389 calories
kilogram per cubic meter	density	kg m^{-3}	1 kg m^{-3} = 0.0624 lb. ft^{-3}
lumen	luminous flux	lm	
lux	illuminance	lx	
meter per second	speed	m s^{-1}	1 m s^{-1} = 3.281 ft s^{-1}

UNIT	QUANTITY	SYMBOL	CONVERSION
meter per second squared	acceleration	$m\ s^{-2}$	
mole per cubic meter	concentration	$mol\ m^{-3}$	
newton	force	N	1 N = 7.218 lb. force
ohm	electric resistance	Ω	
pascal	pressure	Pa	$1\ Pa = \frac{0.145\ lb}{in^{-2}}$
radian per second	angular velocity	$rad\ s^{-1}$	
radian per second squared	angular acceleration	$rad\ s^{-2}$	
square meter	area	m^2	$1\ m^2 = 1.196\ yards^2$
tesla	magnetic flux density	T	
volt	electromotive force	V	
watt	power	W	$1W = 3.412\ Btu\ h^{-1}$
weber	magnetic flux	Wb	

PREFIXES USED WITH SI UNITS		
PREFIX	SYMBOL	VALUE
atto	a	$\times 10^{-18}$
femto	f	$\times 10^{-15}$
pico	p	$\times 10^{-12}$
nano	n	$\times 10^{-9}$
micro	μ	$\times 10^{-6}$
milli	m	$\times 10^{-3}$
centi	c	$\times 10^{-2}$
deci	d	$\times 10^{-1}$
deca	da	$\times 10$
hecto	h	$\times 10^{2}$
kilo	k	$\times 10^{3}$
mega	M	$\times 10^{6}$
giga	G	$\times 10^{9}$
tera	T	$\times 10^{12}$

List of Acronyms

CP	Chemically peculiar
DU	Depleted uranium
GSI	Center for Heavy Ion Research, Darmstadt, Germany
HILAC	Heavy-ion linear accelerator
IUPAC	International Union of Pure and Applied Chemistry
JINR	Joint Institute for Nuclear Research
LBNL	Lawrence Berkeley National Laboratory
LLNL	Lawrence Livermore National Laboratory
MeV	Million electron volts
NATO	North Atlantic Treaty Organization
NIST	National Institute of Standards and Technology
RIKEN	Rikagaku Kenkyūsho—The Institute of Physical and Chemical Research, Japan
UCLA	University of California, Los Angeles

Periodic Table of the Elements

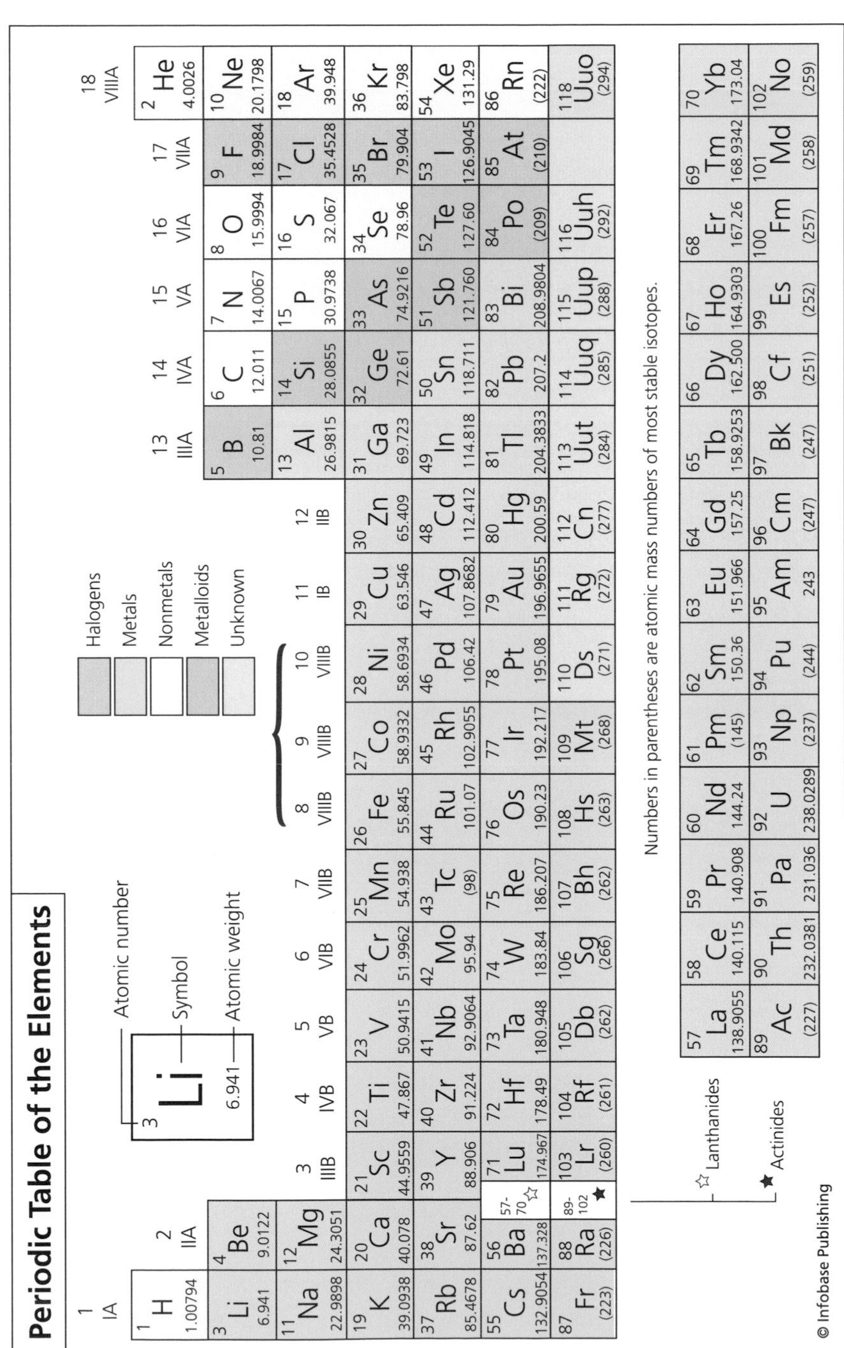

Periodic Table of the Elements

1 IA	2 IIA	3 IIIB	4 IVB	5 VB	6 VIB	7 VIIB	8 VIIIB	9 VIIIB	10 VIIIB	11 IB	12 IIB	13 IIIA	14 IVA	15 VA	16 VIA	17 VIIA	18 VIIIA
1 H 1.00794																	2 He 4.0026
3 Li 6.941	4 Be 9.0122											5 B 10.81	6 C 12.011	7 N 14.0067	8 O 15.9994	9 F 18.9984	10 Ne 20.1798
11 Na 22.9898	12 Mg 24.3051											13 Al 26.9815	14 Si 28.0855	15 P 30.9738	16 S 32.067	17 Cl 35.4528	18 Ar 39.948
19 K 39.0938	20 Ca 40.078	21 Sc 44.9559	22 Ti 47.867	23 V 50.9415	24 Cr 51.9962	25 Mn 54.938	26 Fe 55.845	27 Co 58.9332	28 Ni 58.6934	29 Cu 63.546	30 Zn 65.409	31 Ga 69.723	32 Ge 72.61	33 As 74.9216	34 Se 78.96	35 Br 79.904	36 Kr 83.798
37 Rb 85.4678	38 Sr 87.62	39 Y 88.906	40 Zr 91.224	41 Nb 92.9064	42 Mo 95.94	43 Tc (98)	44 Ru 101.07	45 Rh 102.9055	46 Pd 106.42	47 Ag 107.8682	48 Cd 112.412	49 In 114.818	50 Sn 118.711	51 Sb 121.760	52 Te 127.60	53 I 126.9045	54 Xe 131.29
55 Cs 132.9054	56 Ba 137.328	57-70 ☆ / 71 Lu 174.967	72 Hf 178.49	73 Ta 180.948	74 W 183.84	75 Re 186.207	76 Os 190.23	77 Ir 192.217	78 Pt 195.08	79 Au 196.9655	80 Hg 200.59	81 Tl 204.3833	82 Pb 207.2	83 Bi 208.9804	84 Po (209)	85 At (210)	86 Rn (222)
87 Fr (223)	88 Ra (226)	89-102 ★ / 103 Lr (260)	104 Rf (261)	105 Db (262)	106 Sg (266)	107 Bh (262)	108 Hs (263)	109 Mt (268)	110 Ds (271)	111 Rg (272)	112 Cn (277)	113 Uut (284)	114 Uuq (285)	115 Uup (288)	116 Uuh (292)		118 Uuo (294)

☆ Lanthanides	57 La 138.9055	58 Ce 140.115	59 Pr 140.908	60 Nd 144.24	61 Pm (145)	62 Sm 150.36	63 Eu 151.966	64 Gd 157.25	65 Tb 158.9253	66 Dy 162.500	67 Ho 164.9303	68 Er 167.26	69 Tm 168.9342	70 Yb 173.04
★ Actinides	89 Ac (227)	90 Th 232.0381	91 Pa 231.036	92 U 238.0289	93 Np (237)	94 Pu (244)	95 Am 243	96 Cm (247)	97 Bk (247)	98 Cf (251)	99 Es (252)	100 Fm (257)	101 Md (258)	102 No (259)

Numbers in parentheses are atomic mass numbers of most stable isotopes.

Table of Element Categories

Element Categories

Nonmetals

No.	Symbol	Element
1	H	Hydrogen
6	C	Carbon
7	N	Nitrogen
8	O	Oxygen
15	P	Phosphorus
16	S	Sulfur
34	Se	Selenium

Halogens

No.	Symbol	Element
9	F	Fluorine
17	Cl	Chlorine
35	Br	Bromine
53	I	Iodine
85	At	Astatine

Noble Gases

No.	Symbol	Element
2	He	Helium
10	Ne	Neon
18	Ar	Argon
36	Kr	Krypton
54	Xe	Xenon
86	Ra	Radon

Metalloids

No.	Symbol	Element
5	B	Boron
14	Si	Silicon
32	Ge	Germanium
33	As	Arsenic
51	Sb	Antimony
52	Te	Tellurium
84	Po	Polonium

Alkali Metals

No.	Symbol	Element
3	Li	Lithium
11	Na	Sodium
19	K	Potassium
37	Rb	Rubidium
55	Cs	Cesium
87	Fr	Francium

Alkaline Earth Metals

No.	Symbol	Element
4	Be	Beryllium
12	Mg	Magnesium
20	Ca	Calcium
38	Sr	Strontium
56	Ba	Barium
88	Ra	Radium

Post-Transition Metals

No.	Symbol	Element
13	Al	Aluminum
31	Ga	Gallium
49	In	Indium
50	Sn	Tin
81	Tl	Thallium
82	Pb	Lead
83	Bi	Bismuth

Transactinides

No.	Symbol	Element
104	Rf	Rutherfordium
105	Db	Dubnium
106	Sg	Seaborgium
107	Bh	Bohrium
108	Hs	Hassium
109	Mt	Meitnerium
110	Ds	Darmstadtium
111	Rg	Roentgenium
112	Cn	Copernicium
113	Uut	Ununtrium
114	Uuq	Ununquadium
115	Uup	Ununpentium
116	Uuh	Ununhexium
118	Uuo	Ununoctium

Transition Metals

No.	Symbol	Element	No.	Symbol	Element	No.	Symbol	Element
21	Sc	Scandium	39	Y	Yttrium	72	Hf	Hafnium
22	Ti	Titanium	40	Zr	Zirconium	73	Ta	Tantalum
23	V	Vanadium	41	Nb	Niobium	74	W	Tungsten
24	Cr	Chromium	42	Mo	Molybdenum	75	Re	Rhenium
25	Mn	Manganese	43	Tc	Technetium	76	Os	Osmium
26	Fe	Iron	44	Ru	Ruthenium	77	Ir	Iridium
27	Co	Cobalt	45	Rh	Rhodium	78	Pt	Platinum
28	Ni	Nickel	46	Pd	Palladium	79	Au	Gold
29	Cu	Copper	47	Ag	Silver	80	Hg	Mercury
30	Zn	Zinc	48	Cd	Cadmium			

Lanthanides

No.	Symbol	Element	No.	Symbol	Element	No.	Symbol	Element
57	La	Lanthanum	62	Sm	Samarium	67	Ho	Holmium
58	Ce	Cerium	63	Eu	Europium	68	Er	Erbium
59	Pr	Praseodymium	64	Gd	Gadolinium	69	Tm	Thulium
60	Nd	Neodymium	65	Tb	Terbium	70	Yb	Ytterbium
61	Pm	Promethium	66	Dy	Dysprosium	71	Lu	Lutetium

Actinides

No.	Symbol	Element	No.	Symbol	Element	No.	Symbol	Element
89	Ac	Actinium	94	Pu	Plutonium	99	Es	Einsteinium
90	Th	Thorium	95	Am	Americium	100	Fm	Fermium
91	Pa	Protactinium	96	Cm	Curium	101	Md	Mendelevium
92	U	Uranium	97	Bk	Berkelium	102	No	Nobelium
93	Np	Neptunium	98	Cf	Californium	103	Lr	Lawrencium

Note: The organization of periodic table of the elements, while useful to chemists and physicists, may be confusing to nonscientists in that some groupings of similar elements appear as vertical columns (halogens, for example), some as horizontal rows (lanthanides, for example), and some as a combination of both (nonmetals).

The table of element categories is intended as a quick reference sheet to easily determine which elements belong to which groups. (Element 117 does not appear in this list because it is undiscovered as of the publishing of this book.)

Chronology

1722 The Swedish chemist Axel Fredrik Cronstedt is born on December 23 in Södermanland, Sweden.

1743 The German chemist Martin Heinrich Klaproth is born on December 1 in Wernigerode, Harz, Germany.

1751 Axel Cronstedt discovers a mineral, later called *cerite,* that was found to contain several of the rare earth elements.

1757 The Swedish chemist and army lieutenant Carl Axel Arrhenius is born in Stockholm, Sweden.

1758 The Swedish chemist and mineralogist Bengt Reinhold Geijer is born on November 11.

1760 The Finnish chemist Johan Gadolin is born on June 5 in Åbo, Finland.

1765 Axel Cronstedt dies on August 19 in Stockholm, Sweden.

1766 The Swedish chemist Wilhelm Hisinger is born on December 23 in Vestmanland, Sweden.

1779 The Swedish chemist Jöns Jacob Berzelius is born on August 20 in Väversunda, Sweden.

1787 Carl Arrhenius discovers the mineral ytterite in a quarry near the town of Ytterby on the Swedish island of Resarö. (The name *ytterite* was later changed to *gadolinite.*)

1788 Bengt Geijer publishes the first description of gadolinite.

1789 Martin Klaproth announces the discovery of element 92, uranium, at Berlin, Germany.

1794 Johan Gadolin discovers the compound yttria in gadolinite; element 39, yttrium, was later isolated from yttria.

1797 The Swedish chemist, mineralogist, and army surgeon Carl Gustav Mosander is born on September 10 in Kalmar, Sweden.

1803 Jöns Jacob Berzelius and Wilhelm Hisinger announce their discovery of ceria, an oxide of cerium, at Vestmanland, Sweden. Martin Klaproth independently announces his discovery of ceria.

1811 The French chemist Eugène-Melchior Péligot is born in Paris, France.

The German chemist Robert Wilhelm Bunsen is born on March 31 in Göttingen, Germany.

1815 Bengt Geijer dies on his birthday, November 11 in Sweden.

Jöns Jacob Berzelius announces the discovery of element 90, thorium, at Stockholm, Sweden.

1817 Martin Klaproth dies on January 1 in Berlin, Germany.

The Swiss chemist Jean-Charles Galissard de Marignac is born on April 24 in Geneva, Switzerland.

1818 The American chemist John Lawrence Smith is born on December 17 near Charleston, South Carolina.

1824 The German physicist Gustav Kirchhoff is born on March 12 in Königsberg, East Prussia.

1827 The Swiss chemist Jacques-Louis Soret is born on June 30 in Geneva, Switzerland.

1833 The Swedish chemist and inventor Alfred Nobel is born on October 21 in Stockholm, Sweden.

1838 The French Paul-Émile Lecoq de Boisbaudran is born on April 18 in Cognac, France.

1839 Carl Mosander announces the discovery at Stockholm, Sweden, of a rare earth oxide that he names *lanthana*. Element 57, lanthanum, is later obtained from lanthana.

1840 The Swedish chemist and geologist Per Teodor Cleve is born on February 10 in Stockholm.

The Swedish chemist Lars Fredrik Nilson is born on May 27 in Östergötland, Sweden.

1841 Carl Mosander isolates a new rare earth oxide he calls *didymium,* which was shown later to be a mixture of element 59, praseodymium, and element 60, neodymium.

Eugène-Melchior Péligot announces the first isolation of pure uranium at Paris, France.

1843 Carl Mosander analyzes yttria, isolates element 39, yttrium, and discovers two new rare earth oxides that he calls *erbia* and *terbia.*

1848 Jöns Jacob Berzelius dies on August 7 in Stockholm, Sweden.

1851 The American chemist Thomas Herbert Norton is born on June 30 in Rushford, New York.

1852 The German chemist Friedrich Oskar Giesel is born on May 20 in Winsko, Poland.

Wilhelm Hisinger dies on June 28 in Vestmanland, Sweden.

Johan Gadolin dies on August 15 in Wirmo, Finland.

The French chemist Eugène-Anatole Demarçay is born in Paris, France.

The French physicist Antoine Henri Becquerel is born on December 15 in Paris, France.

1853 The American chemist William Francis Hillebrand is born on December 12 in Honolulu, Hawaii.

1855 The Czech chemist Bohuslav Brauner is born on May 8.

Robert Bunsen develops the laboratory burner known today as the Bunsen burner.

1858 Carl Mosander dies on October 15 in Ångsholm near Drottningholm, Sweden.

The Austrian scientist Carl Auer von Welsbach is born on September 1 in Vienna, Austria.

1859 Robert Bunsen and Gustav Kirchhoff begin their collaboration that results in the development of emission spectroscopy as an essential analytical tool.

1867 The Polish-French chemist Marie Skłodowska Curie is born on November 7 in Warsaw, Poland.

Alfred Nobel files his patent for the invention of dynamite.

1871 The British physicist Ernest Rutherford is born on August 30 in Brightwater, New Zealand.

1872 The French chemist Georges Urbain is born on April 12 in Paris, France.

1874 John Lawrence Smith serves as president of the American Association for the Advancement of Science.

The French chemist André-Louis Debierne is born on July 14 in Paris, France.

1875 William Hillebrand and Thomas Norton isolate the first pure sample of cerium at the National Bureau of Standards in Washington, D.C.

1877 John Lawrence Smith serves as president of the American Chemical Society.

The English chemist Frederick Soddy is born on September 2 in Eastbourne, England.

1878 Jean-Charles de Marignac announces the discovery of element 70, ytterbium.

The Austrian physicist Lise Meitner is born on November 17 in Vienna, Austria.

1879 The German chemist Otto Hahn is born on March 8 in Frankfurt-am-Main, Germany.

The discovery of element 67, holmium, is announced by Per Cleve at Uppsala, Sweden, and independently by Marc Delafontaine and Jacques-Louis Soret at Geneva, Switzerland.

Per Cleve announces the discovery of element 69, thulium, at Uppsala, Sweden.

Lars Nilson announces the discovery of element 21, scandium, in Sweden.

Paul-Émile Lecoq de Boisbaudran announces the discovery of element 62, samarium, at Paris, France.

1880 Jean-Charles de Marignac observes spectral lines attributed to element 64, gadolinium, at Geneva, Switzerland.

The British-American chemist Charles James is born in Northhamptonshire, England.

1882 Bohuslav Brauner identifies spectral lines later attributed to element 59, praseodymium, and element 60, neodymium.

1883 John Lawrence Smith dies on October 12 in Louisville, Kentucky.

1885 Carl Auer von Welsbach isolates neodymium and praseodymium at Vienna, Austria.

1886 Paul-Émile Lecoq de Boisbaudran announces the discovery of element 66, dysprosium, at Paris, France. Boisbaudran also isolates gadolinium.

1887 The Polish-American chemist Kazimierz Fajans is born on May 27 in Warsaw, Poland.

Gustav Kirchhoff dies on October 17 in Berlin, Germany.

English physicist Henry Moseley is born on November 23 in Weymouth, England.

1889 The British chemist Alexander Fleck is born on November 11 in Glasgow, Scotland.

1890 Jacques-Louis Soret dies on May 13 in Geneva, Switzerland.

Eugène-Melchior Péligot dies in Paris, France.

1891 The English physicist James Chadwick is born on October 20 in Bollington, Cheshire, England.

1894 Jean-Charles Marignac dies on April 15 in Geneva, Switzerland.

1896 Henri Becquerel announces his discovery of radioactivity.

Alfred Nobel dies on December 10 in San Remo, Italy.

1899 André-Louis Debierne announces his discovery of element 89, actinium, at Paris, France.

Lars Nilson dies on May 14.

Robert Bunsen dies on August 16 in Heidelberg, Germany.

1901 Eugène-Anatole Demarçay announces the discovery of element 63, europium, at Paris, France.

The American physicist Ernest Orlando Lawrence is born on August 8 in Canton, South Dakota.

The Italian physicist Enrico Fermi is born on September 29 in Rome, Italy.

The first Nobel Prizes are awarded in December.

1902 The German chemist Friedrich Wilhelm "Fritz" Strassmann is born on February 22 in Boppard, Germany.

Friedrich Giesel announces his independent discovery of actinium.

1903 Pierre and Marie Curie, together with Henri Becquerel, share the Nobel Prize in physics.

1904 Eugène-Anatole Demarçay dies in December in France.

1905 The Italian physicist Emilio Segrè is born on February 1 in Tivoli, Italy.

1907 The American physicist Edwin Mattison McMillan is born on September 18 in Redondo Beach, California.

Ernest Rutherford discovers the atomic nucleus.

The discovery of element 71, lutetium, is announced by Georges Urbain at Paris, France, and independently by Carl Auer von Welsbach in Austria, and by Charles James at Durham, New Hampshire.

1908 Ernest Rutherford receives the Nobel Prize in chemistry for his work on radioactivity and the chemistry of radioactive substances.

Henri Becquerel dies on August 25 in Le Croisic, Brittany, France.

1911 Marc Delafontaine dies.

1912 The American chemist Glenn Theodore Seaborg is born on April 19 in Ishpeming, Michigan.

Paul-Émile Lecoq de Boisbaudran dies on May 28 in Paris, France.

The American chemist Charles Dubois Coryell is born on February 21 near Los Angeles, California.

1913 Russian physicist Georgiĭ Nikolaevich Flerov is born on March 2 in Rustov-on-Don, Russia.

The American physicist Phillip Hauge Abelson is born on April 27 in Tacoma, Washington.

Henry Moseley discovers the importance of nuclear charge in identifying an element's atomic number.

Frederick Soddy develops the concept of isotopes.

1914 Ernest Rutherford is knighted by King George V.

1915 The American chemist Isadore Perlman is born on April 12 in Milwaukee, Wisconsin.

The American nuclear scientist Albert Ghiorso is born on July 15 in Vallejo, California.

Henry Moseley is killed in Turkey on August 10.

1917 The discovery of element 91, protactinium, is announced independently by Otto Hahn and Lise Meitner at Berlin, Germany; by Kazimierz Fajans at Karlsruhe, Germany; and by Frederick Soddy, John Arnold Cranston, and Alexander Fleck at Glasgow, Scotland.

1918 The American chemist Jacob A. Marinsky is born in Buffalo, New York.

The American chemist Lawrence Elgin Glendenin is born on November 8 in Bay City, Michigan.

1919 Ernest Rutherford performs the first transmutation of one element into another element.

1920 The American chemist Kenneth Street, Jr., is born on January 30 in Berkeley, California.

The American physicist Owen Chamberlain is born on July 10 in San Francisco, California.

1921 Frederick Soddy is awarded the Nobel Prize in chemistry for his discovery of isotopes.

1923 The Norwegian chemist Torbjørn Sikkeland is born on August 3.

1925 William Hillebrand dies on February 7.

1927 The American chemist Gregory R. Choppin is born in Texas.

Friedrich Giesel dies on November 14 in Brunswick, Germany.

1928 Charles James dies on December 10 in Durham, New Hampshire.

1929 Ernest Orlando Lawrence develops the first cyclotron at the University of California, Berkeley.

Carl Auer von Welsbach dies on August 4 in Treibach, Austria.

1932 James Chadwick discovers the neutron.

1934 The German chemists Wilhelm Klemm and Heinrich Bommer produce the first samples of pure erbium metal.

Marie Curie dies on July 4 in Passy, France.

1935 Bohuslav Brauner dies on February 15 in Prague, Czechoslovakia.

James Chadwick receives the Nobel Prize in physics for his discovery of the neutron.

1938 Lise Meitner, Otto Frisch, Otto Hahn, and Fritz Strassmann announce their discovery of nuclear fission based on work done at Berlin, Germany.

Georges Urbain dies on November 5 in Paris, France.

1939 Emilio Segrè and Carlo Perrier announce their discovery of element 43, technetium.

Ernest Orlando Lawrence receives the Nobel Prize in physics for the invention of the cyclotron.

1940 The German physicist Gottfried Münzenberg is born on March 17 in Nordhausen, Germany.

Edwin McMillan and Phillip Abelson announce their discovery of element 93, neptunium, at Berkeley, California.

Emilio Segrè, Dale R. Corson, and Kenneth R. Mackenzie announce their discovery of element 85, astatine.

Glenn Seaborg, Arthur Wahl, and Joseph Kennedy announce their discovery of element 94, plutonium, at Berkeley, California.

1942 Glenn T. Seaborg, Burris B. Cunningham, and Louis B. Werner isolate the first visible sample of plutonium on August 20.

Enrico Fermi's group demonstrates the first controlled nuclear chain reaction on December 2 in Chicago, Illinois.

1944 Otto Hahn receives the Nobel Prize in chemistry for the discovery of nuclear fission.

The German chemist Sigurd Hofmann is born.

Glenn Seaborg, Albert Ghiorso, Ralph A. James, and Leon O. Morgan discover element 95, americium, at the University of Chicago, Illinois. Seaborg, Ghiorso, and Morgan also discover element 96, curium. (The public announcement of the discoveries of the two elements occurs in 1945 after the conclusion of World War II.)

1945 Jacob Marinsky, Lawrence Glendenin, and Charles Coryell announce their discovery of element 61, promethium, at Oak Ridge National Laboratory, Tennessee.

James Chadwick is knighted by King George VI.

The American scientists demonstrate the first successful nuclear explosion on July 16 at the Trinity Test Site in New Mexico.

The United States drops uranium fission bombs on Hiroshima and Nagasaki, Japan, on August 6 and 9, respectively.

1947 Louis B. Werner and Isadore Perlman isolate the first visible amounts of curium at Berkeley, California.

1949 André-Louis Debierne dies on August 31 in Paris, France.

Glenn Seaborg, Albert Ghiorso, and Stanley G. Thompson announce their discovery of element 97, berkelium, at Berkeley, California.

1950 Glenn Seaborg, Albert Ghiorso, Kenneth Street, Jr., and Stanley G. Thompson announce their discovery of element 98, californium, at Berkeley, California.

1951 Edwin McMillan and Glenn Seaborg share the Nobel Prize in chemistry for their discoveries in the chemistry of transuranium elements.

1952 Glenn Seaborg, Albert Ghiorso, Stanley G. Thompson, Gregory R. Choppin, and Bernard G. Harvey announce their discoveries of element 99, einsteinium, and element100, fermium, from the debris of a thermonuclear explosion in the Pacific basin.

1954 Enrico Fermi dies on November 28 in Chicago, Illinois.

1955 Glenn Seaborg, Albert Ghiorso, Stanley G. Thompson, Gregory R. Choppin, and Bernard G. Harvey announce their discovery of element 101, mendelevium, at Berkeley, California.

Emilio Segrè and Owen Chamberlain announce their discovery of the antiproton at Berkeley, California.

1956 Frederick Soddy dies on September 22 in Brighton, England.

1957 Reports of the synthesis of element 102, nobelium, are reported from the Nobel Institute in Stockholm, Sweden. (This claim is later retracted.)

Ernest Orlando Lawrence receives the Enrico Fermi Award from the U.S. Atomic Energy Commission.

1958 Glenn Seaborg, Albert Ghiorso, John R. Walton, and Torbjørn Sikkeland announce their positive identification of nobelium at Berkeley, California.

A team led by Georgiĭ Nikolaevich Flerov at Dubna in the Soviet Union provides additional confirmation of nobelium.

Ernest Orlando Lawrence dies on August 27 in Palo Alto, California. The Regents of the University of California vote to rename the Berkeley and Livermore Radiation Laboratories the Lawrence Berkeley Laboratory and the Lawrence Livermore Laboratory in his honor.

1959 Glenn T. Seaborg receives the Enrico Fermi Award from the U.S. Atomic Energy Commission.

1961 Albert Ghiorso, Torbjørn Sikkeland, Almon Larsh, and Robert M. Latimer announce their discovery of element 103, lawrencium, at Berkeley, California.

1964 Scientists at the Joint Institute for Nuclear Research at Dubna, Russia, give the first report of synthesizing element 104, *rutherfordium.*

1966 Lise Meitner, Otto Hahn, and Fritz Strassmann receive the Presidential Enrico Fermi Award from the U.S. government.

1968 Otto Hahn dies on July 28 in Göttingen, Germany.

Alexander Fleck dies on August 6 in Great Britain.

Lise Meitner dies on October 27 in Cambridge, England.

Georgiĭ Nikolaevich Flerov at the Joint Institute for Nuclear Research at Dubna, Russia, reports the synthesis of element 105, first named *hahnium,* but which name was later changed to *dubnium.*

1969 The American scientists at the Lawrence Berkeley Laboratory in California fail to duplicate the Russians' 1966 synthesis of rutherfordium, but succeed in producing it by a different synthesis.

1970 Scientists at Berkeley report the synthesis of element 105 by a different reaction than the one reported in Dubna in 1968.

1971 Charles Coryell dies on January 7 in Massachusetts.

Peter Armbruster becomes the scientific leader of the divisions of Atomic Physics and Nuclear Chemistry at GSI in Darmstadt, Germany.

1972 John Arnold Cranston dies on April 25.

1974 Scientists from the Lawrence Berkeley Laboratory and the Lawrence Livermore Laboratory (both in California) report the synthesis of element 106, seaborgium. The claim is not substantiated until 1993.

James Chadwick dies on July 24 in Cambridge, England.

1975 Kazimierz Fajans dies on May 18 in Ann Arbor, Michigan.

1976 Glenn T. Seaborg serves as president of the American Chemical Society.

Stanley G. Thompson dies on July 16 in Berkeley, California.

Scientists at Dubna report the first synthesis of element 107, bohrium, a claim that remained disputed.

1979 Glenn Seaborg receives the Priestley Medal from the American Chemical Society.

1980 Fritz Strassmann dies on April 22 in Mainz, Germany.

1981 Peter Armbruster and Gottfried Münzenber at GSI in Darmstadt, Germany, report the positive discovery of element 107.

1982 Peter Armbruster and Gottfried Münzenber at GSI in Darmstadt, Germany, report the discovery of element 109, meitnerium.

1984 Peter Armbruster and Gottfried Münzenber at GSI in Darmstadt, Germany, report the first synthesis of element 108, hassium.

1989 Emilio Segrè dies on April 22 in Lafayette, California.

1990 Georgiĭ Nikolaevich Flerov dies on November 19 in Russia.

1991 Isadore Perlman dies on August 3 in Los Alamitos, California.

Edwin McMillan dies on September 7 in El Cerrito, California.

The Glenn T. Seaborg Institute for Transactinium Science is established at Lawrence Livermore Laboratory, California.

1993 A team from the Lawrence Berkeley Laboratory led by Darleane Hoffman and Kenneth Gregorich confirms the original discovery of element 106.

1994 Sigurd Hofmann, Victor Ninov, Fritz Hessberger, Helmut Folger, Hans-Joachim Schött, Peter Armbruster, Gottfried Münzenber, and coworkers at GSI in Darmstadt, Germany, announce their discoveries of element 110, darmstadtium, and element 111, roentgenium.

1996 Sigurd Hofmann, Victor Ninov, Fritz Hessberger, Helmut Folger, Peter Armbruster, Gottfried Münzenber, and coworkers at GSI in Darmstadt, Germany, announce their discovery of element 112 (later named *copernicium*).

1997 After several years of controversy, the International Union for Pure and Applied Chemistry approves the name *seaborgium* for element 106.

1998 Scientists at the Joint Institute for Nuclear Research in Dubna, Russia, report the first synthesis of element 114 (as yet unnamed).

The Glenn T. Seaborg Center for Teaching and Learning Science and Mathematics and the new Seaborg Science Complex are established at Northern Michigan University in Marquette.

Mountain Pass mine in California ceases rare earth mining operations due to environmental issues and competition from Bayun Obo mine in China.

1999 Glenn Seaborg dies on February 25 in Lafayette, California.

2000 Scientists at the Joint Institute for Nuclear Research in Dubna, Russia, report the first synthesis of element 116 (as yet unnamed).

The American chemist Darleane C. Hoffman receives the Priestley Medal from the American Chemical Society.

2001 Scientists at the Lawrence Berkeley Laboratory in California report the syntheses of elements 116 and 118. These claims were later retracted.

2004 Phillip Abelson dies on August 1 in Bethesda, Maryland.

Albert Ghiorso receives the Lifetime Achievement Award of the Radiochemistry Society.

Scientists at the Lawrence Livermore National Laboratory in California and at the Joint Institute for Nuclear Research in Dubna, Russia, report the first syntheses of elements 113 and 115 (as yet unnamed).

2005 Jacob Marinsky dies on September 1 in Buffalo, New York.

2006 Owen Chamberlain dies on February 28 in Berkeley, California.

Kenneth Street, Jr., dies on March 13 in California.

Scientists at the Lawrence Livermore National Laboratory in California and the Joint Institute for Nuclear Research in Dubna, Russia, report the first syntheses of elements 116 and 118 (as yet unnamed).

2008 Lawrence Glendenin dies on November 22.

2009 The Germans' claim of having discovered element 112 is confirmed. The name *copernicium* is proposed.

On July 22, U.S. Secretary of the Interior Ken Salazar announces a two-year moratorium on new uranium claims in and around Grand Canyon National Park.

The United Nations Security Council votes on September 25 to work toward a planet free from nuclear weapons.

2010 The IUPAC officially adopts the name *copernicium* for element 112 and assigns it the symbol Cn.

A team of Russian and U.S. scientists synthesize the first atoms of element 117 at JINR in Dubna, Russia.

On April 8, the presidents of the United States and Russia sign an agreement to reduce stockpiles by 30 percent.

Glossary

acid a type of compound that contains hydrogen and dissociates in water to produce hydrogen ions.

acid mine drainage acidic water that drains from mining facilities into the surrounding soil and aquifers.

actinide concept the recognition in the 1940s that elements 89 through 92 (and subsequent elements with higher atomic numbers) should be displayed in their own row below the periodic table in analogy to the lanthanides.

actinides the elements ranging from thorium (atomic number 90) to lawrencium (number 103); all have two outer electrons in the "7s" subshell plus an increasing number of electrons in the "5f" subshell.

alkali metals the elements in column IA of the periodic table (exclusive of hydrogen); they all are characterized by a single valence electron in an "s" subshell.

alkaline earth metals the elements in column IIA of the periodic table; all are characterized by two valence electrons that fill an "s" subshell.

alpha decay a mode of radioactive decay in which an alpha particle is emitted. The daughter isotope has an atomic number two units less than the atomic number of the parent isotope, and a mass number that is four units less.

alpha particle a nucleus of helium 4.

alternating current electric current that has the property of reversing direction periodically in a circuit; the basic mode of current used in distributed power systems worldwide.

analytical chemistry the branch of chemistry that determines the chemical constitution of a chemical sample. (See **qualitative analysis** and **quantitative analysis.**)

angular momentum the product of a particle's mass, velocity, and its radius of curvature about an axis of rotation; the different subshells in which electrons are found are described by the electrons' angular momentum.

anion an atom with one or more extra electrons giving it a net negative charge.

antineutrino the antiparticle of a neutrino.

antiproton the antiparticle of a proton; protons have positive charges while antiprotons have negative charges.

antiseptic able to clean or disinfect; literally "against *sepsis*."

anode the site of oxidation in an electrochemical cell. In an electrolytic cell, the anode is positively charged; in a galvanic cell, the anode is negatively charged.

aqueous describing a solution in water.

aquifer an underground body of water surrounded by rock or soil.

atom the smallest part of an element that retains the element's chemical properties; atoms consist of protons, neutrons, and electrons.

atomic mass the mass of a given isotope of an element—the combined masses of all its protons, neutrons, and electrons.

atomic number the number of protons in an atom of an element; the atomic number establishes the identify of an element.

atomic weight the mean weight of the atomic masses of all the atoms of an element found in a given sample, weighted by isotopic abundance.

base a substance that reacts with an acid to give water and a salt; a substance that, when dissolved in water, produces hydroxide ions.

beta decay a mode of radioactive decay in which a beta particle—an ordinary electron—is emitted; the daughter isotope has an atomic number one unit greater than the atomic number of the parent isotope, but the same mass number.

beta particle an electron that results from radioactive decay; symbol $= {}^{0}_{-1}e$ if the particle is an ordinary electron with a negative charge, symbol $= {}^{0}_{+1}e$ if it is a positron with a positive charge.

carbon dating a method for determining the age of a once-living object by measuring the ratio of carbon 14 to carbon 12 in the sample.

carcinogenic cancer-causing.

catalyst a chemical substance that speeds up a chemical reaction without itself being consumed by the reaction.

catalytic converter a device in motor vehicles that uses a catalyst—usually platinum or another platinum group metal—to reduce emissions of pollutant gases.

cathode the site of reduction in an electrochemical cell. In an electrolytic cell, the cathode is negatively charged; in a galvanic cell, the cathode is positively charged.

cation an atom that has lost one or more electrons to acquire a net positive charge.

ceria subgroup the lanthanides that tend to be found in ores with cerium.

chain reaction the self-sustaining process in which the production and capture of neutrons causes a cascading number of nuclear fissions to occur in adjacent nuclei (usually of uranium or plutonium).

chemical bond the force of attraction holding atoms together in molecules or crystals.

chemical change a change in which one or more chemical elements or compounds form new compounds; in a chemical change, the names of the compounds change.

chemically peculiar star any star that shows anomalies in the expected abundance of various elements, usually having a high proportion of heavy elements as compared to solar abundance.

chromatography a method for separating and analyzing the components of mixtures of liquids, gases, or solutions.

cold war the period starting just after World War II and lasting until about 1990, when tensions between the United States and the USSR were such that only competing nuclear arsenals were seen to keep the peace.

complex ion any ion that contains more than one atom.

compound a pure chemical substance consisting of two or more elements in fixed, or definite, proportions.

contraction a decrease in size; see **lanthanide contraction.**

control rod a rod used in nuclear reactors to regulate the rate of the fission process by absorbing neutrons.

core electrons the electrons located in shells close to an atom's nucleus; core electrons usually are not involved in chemical bonding.

corrosion degradation of a metal surface by a chemical or electrochemical reaction.

cosmochronology the method of using ratios of radioactive elements in the universe to determine the age of astrophysical objects and systems.

cosmochronometer a ratio of radioactive elements that can be used to determine the age of astrophysical objects.

covalent bond a chemical bond formed by sharing valence electrons between two atoms (in contrast to an ionic bond, in which one or more valence electrons are transferred from one atom to another atom).

criticality the condition in which a nuclear chain reaction is self-sustaining.

critical point temperature the temperature of a pure substance at which the liquid and gaseous phases become indistinguishable.

cross section a measure of the probability that a projectile particle will collide with a target particle.

crystallization the process of forming crystals from liquids or solids.

cyclotron an apparatus that accelerates charged particles to very high energies as the particles travel in circular paths.

daughter isotope an isotope produced by the radioactive decay of another (parent) isotope.

decant to pour a liquid from one container into another container without disturbing any solid that may be present in the first container.

deuterium an atom of heavy hydrogen that contains one proton, one neutron, and one electron.

deuteron a nucleus of heavy hydrogen that contains one proton and one neutron.

DNA deoxyribonucleic acid; the genetic material of all living organisms that is found in the organisms' chromosomes.

doubling time the time it takes for the number of items in a sample to double.

earth the oxide of an element. Few elements are extracted from the ground in pure form; most occur as minerals such as oxides.

eka- Mendeleev's prefix for an unknown element he predicted would have properties similar to the known element located above it in the period table. Examples: gallium = eka-aluminum; germanium = eka-silicon; scandium = eka-boron.

electrolysis the production of a chemical reaction achieved by passing an electrical current through an ionic solution.

electrolytic related to the use of an electrolyte.

electron a subatomic particle found in all neutral atoms and negative ions; possesses the negative charges in atoms.

electronic configuration a shorthand notation that indicates which atomic orbitals of an element's atoms are occupied by electrons and in what arrangement.

electrostatic the type of interaction that exists between electrically charged particles; electrostatic forces attract particles together if the particles have charges of opposite sign, while the forces cause particles that have charges of like sign to repel one another.

element a pure chemical substance that contains only one kind of atom.

elution the process of washing the components of a mixture through a chromatography column; see **ion chromatography.**

emission spectrum the range of wavelengths of radiation that an object produces when the object is heated.

energy a measure of a system's ability to do work; expressed in units of joules.

enrichment an increase in the percentage of one isotope of an element compared to the percentage that occurs naturally.

exponential function a function of the form e^x, where e is 2.7182818281828 . . . and x is a number related to the doubling time of a sample.

fallout the radioactive debris from a nuclear explosion or a nuclear accident.

family see **group.**

f-block the two rows of elements that are located below the periodic table; the first row, the *lanthanides,* have outermost electrons in "4f" subshells; the second row, the *actinides,* have outermost electrons in "5f" subshells.

fermion a class of elementary particles that includes electrons, photons, and neutrons.

Fermi-Dirac statistics a set of mathematical rules that govern the behavior of certain elementary particles now called *fermions.*

ferromagnetic any metallic material that contains tiny magnetic domains that can be aligned by an external magnetic field.

fission see **nuclear fission.**

fission reactor a power source that relies on the fission of nuclei (usually uranium 235) to provide heat to turn turbines.

fluorescence the spontaneous emission of light from atoms or molecules when electrons make transitions from states of higher energy to states of lower energy.

fractional crystallization a technique for separating a mixture of soluble compounds. The mixture is heated and then allowed to cool; the components of the mixture will crystallize at different temperatures.

fusion may refer to a substance's phase change from solid to liquid or to nuclear fusion; see **nuclear fusion.**

galactic halo the region at the outer edges of the Milky Way where very old metal-poor stars can be observed.

gamma decay a mode of radioactive decay in which a very high energy photon of electromagnetic radiation—a gamma ray—is emitted; the daughter isotope has the same atomic number and mass number as the parent isotope, but lower energy.

gamma ray a high-energy photon.

globular cluster a spherical cluster of stars normally found on the outer edges of a galaxy and consisting mostly of very old stars.

group the elements that are located in the same column of the periodic table; also called a family, elements in the same column have similar chemical and physical properties.

half-life the time required for half of the original nuclei in a sample to decay.

halogen the elements in column VIIB of the periodic table; all of them share a common set of seven valence electrons in an nth energy level such that their outermost electronic configuration is ns^2np^5.

heavy hydrogen the isotope of hydrogen with one proton and one neutron in its nucleus; symbol = ${}^{2}_{1}H$.

hexafluoride an inorganic compound containing 6 fluoride ions (F^-).

homologue an element that lies below another element in the periodic table is said to be the homologue of that element.

hydronium ion an ion with the formula H_3O^+.

hydrosphere the watery part of Earth; the oceans, lakes, rivers, icecaps, and glaciers.

inner electrons see **core electrons.**

inner transition elements the lanthanides and actinides.

in situ leaching the process of introducing a solution in which an ore dissolves, and the solution is then pumped out on the far side of the deposit.

inorganic compound any chemical compound that does not contain both carbon and hydrogen.

ion an atom having a net electrical charge.

ion exchange chromatography the separation of different ions achieved by passing an ionic solution through a column containing a solid; different ions are absorbed by the solid and then released again at different rates, resulting in different times at which the ions exit the column.

ionic bond a strong electrostatic attraction between a positive ion and a negative ion that holds the two ions together.

ionizing particle a particle with properties (such a high-energy or electrical charge) that allow it to ionize particles in any medium through which it passes.

ionizing radiation any radiation that can remove an electron from an atom, molecule, or cell.

island of stability on a chart of nuclides on which numbers of protons are graphed in relation to numbers of neutrons, the location of a superheavy element that has a half-life longer than one second.

isotope a form of an element characterized by a specific mass number; the different isotopes of an element have the same number of protons but different numbers of neutrons, hence different mass numbers.

lanthanide contraction the observed decrease in sizes of atoms from lanthanum (element 57) to lutetium (element 71).

lanthanides the elements ranging from cerium (atomic number 58) to lutetium (number 71); they all have two outer electrons in the "6s" subshell plus increasingly more electrons in the "4f" subshell.

lanthanide series see **lanthanides.**

lithium deuteride a compound of lithium and deuterium, used in thermonuclear weapons.

lithosphere the solid part of Earth; depending on the context, may include Earth's crust, mantle, and core.

magic number referring to nuclei that have closed energy shells. Groupings of 2, 8, 20, 50, 28, 82, and 126 make filled nuclear shells.

magnetic flux a measurement of the amount of magnetic field passing through a given area.

main group element an element in one of the first two columns or one of the right-hand six columns of the periodic table; distinguished from transition metals, which are located in the middle of the table, and from rare earths, which are located in the lower two rows shown apart from the rest of the table.

malleability the ability of a substance such as a metal to change shape without breaking; metals that are malleable can be hammered into thin sheets.

Manhattan Project the World War II project that culminated in the development of the atom bomb; named for the Manhattan District of the U.S. Army Corps of Engineers.

mass a measure of an object's resistance to acceleration; determined by the sum of the elementary particles composing the object.

mass number the sum of the number of protons and neutrons in the nucleus of an atom; see **isotope.**

mass spectrometry a process by which a substance is vaporized and ionized so that its different constituent species can be electromagnetically separated for analysis.

metal any of the elements characterized by being good conductors of electricity and heat in the solid state; approximately 75 percent of the elements are metals.

metalloid (also called semimetal) any of the elements intermediate in properties between the metals and nonmetals; the elements in the periodic table located between metals and nonmetals.

metallurgy the branch of engineering that deals with the production of metals from their ores, the manufacture of alloys, and the use of metals in engineering applications.

metastable an atomic state above the ground state that can be populated by the decay of an electron from a higher state. The state is relatively long-lived.

microgram a millionth of a gram, or 10^{-6}g; symbol = μg.

mineralogist a geologist who studies minerals.

mischmetal an alloy comprising a combination of rare earth metals.

mixture a system that contains two or more different chemical substances.

moderator in a nuclear reactor, a medium that slows the speed of initially fast-moving neutrons.

nanometer a metric unit of length equal to 1 billionth (10^{-9}) of a meter; abbreviated "nm."

neutrino an elementary particle that has no charge and that travels at nearly the speed of light.

neutron the electrically neutral particle found in the nuclei of atoms.

neutron capture the process in which a neutron collides with an atomic nucleus and is captured by that nucleus.

noble gas any of the elements located in the last column of the periodic table—usually labeled column VIII or 18, or possibly column 0.

nonmetal the elements on the far right-hand side of the periodic table that are characterized by little or no electrical or thermal conductivity, a dull appearance, and brittleness.

nonproliferation usually referring to nuclear arsenals, the situation in which countries willingly choose or are compelled not to build nuclear weapons.

nuclear fission the process in which certain isotopes of relatively heavy atoms such as uranium or plutonium spontaneously break apart into fragments; accompanied by the release of large amounts of energy.

nuclear fusion the process in which nuclei lighter than iron can combine to form heavier nuclei; accompanied by the release of large amounts of energy.

nuclear medicine the branch of medicine that uses radioactive isotopes for diagnosis or treatment of disease.

nuclear model the model of the atom in which the protons and neutrons, and thus the positive charges and most of the atom's mass, are concentrated in a tiny center called the *nucleus;* the electrons are located outside the nucleus.

nuclear reaction a type of chemical reaction in which there are changes in the composition or energies of atomic nuclei; if the number of protons in an atom changes, also called a *transmutation* reaction.

nucleon a particle found in the nucleus of atoms; a proton or a neutron.

nucleosynthesis the process by which atomic nuclei are synthesized.

nucleus the small, central core of an atom.

nuclide an atomic nucleus characterized by its numbers of protons and neutrons.

ordinary chemical reaction a type of chemical reaction in which the compositions of atomic nuclei do not change; the only changes are typically in numbers and types of chemical bonds that are present.

Most often, the atoms in chemical compounds are rearranged to form different compounds.

oxidation an increase in an atom's oxidation state; accomplished by a loss of electrons or an increase in the number of chemical bonds to atoms of other elements; see **oxidation state.**

oxidation-reduction reaction a chemical reaction in which one element is oxidized and another element is reduced.

oxidation state a description of the number of atoms of other elements to which an atom is bonded. A neutral atom or neutral group of atoms of a single element are defined to be in the zero oxidation state. Otherwise, in compounds, an atom is defined as being in a positive or negative oxidation state depending upon whether the atom is bonded to elements that, respectively, are more or less electronegative than that atom is.

oxidizing agent a chemical reagent that causes an element in another reagent to be oxidized.

oxyanion a negative ion that contains one or more oxygen atoms plus one or more atoms of at least one other element.

oxychloride a negative ion that contains both oxygen and chlorine.

oxyfluoride a negative ion that contains both oxygen and fluorine.

parent isotope an atom that undergoes radioactive decay into a daughter isotope.

particulates matter, usually in air, comprising various-size small particles.

pentavalent referring to an atom that has five chemical bonds.

pentoxide an inorganic chemical compound that contains five oxygen atoms

period any of the rows of the periodic table; rows are referred to as periods because of the periodic, or repetitive, trends in the properties of the elements.

periodic table an arrangement of the chemical elements into rows and columns such that the elements are in order of increasing atomic number, and elements located in the same column have similar chemical and physical properties.

permanent magnet any material that is capable of exerting magnetic force as a result of the magnetic alignment of domains within the material, rather than being the result of a current or external magnetic field.

pH a measure of the acidity of an aqueous solution; low pHs are strongly acid and high pHs are strongly basic.

phase referring to whether a substance is in the solid, liquid, or vapor state; also may refer to the different components of a mixture.

photon the name for the particle nature of light.

physical change any transformation that results in changes in a substance's physical state, such as color, temperature, dimensions, or other physical properties; the chemical identity of the substance remains unchanged in the process.

physical state the condition of a substance being either a solid, liquid, or gas.

post-transition metal a naturally occurring metal located in p-block of the periodic table: aluminum, gallium, indium, tin, thallium, lead, and bismuth.

potential difference the amount of voltage rise or drop between two positions—often gauged between the locations of two electrodes or nodes in a circuit.

precipitants grains of metals that form in alloys.

precipitate in chemistry, a solid that forms by the combination of positive and negative ions. In meteorology, the various forms of water that deposit from the atmosphere.

pressure differential an area of increasing or decreasing pressure.

primordial anything associated with the beginning of time, either on Earth or in the universe.

product the compounds that are formed as the result of a chemical reaction.

proton the positively charged subatomic particle found in the nuclei of atoms.

qualitative analysis the process of identifying what substances are present in a mixture.

quantitative analysis the process of determining how much of a substance is present in a mixture.

quantum a unit of discrete energy on the scale of single atoms, molecules, or photons of light.

radioactive decay the spontaneous disintegration of an atomic nucleus accompanied by the emission of a subatomic particle or gamma ray.

radioactive decay series modes of decay of a particular radioactive series, usually including daughters, half-lives, and other decay products.

radiology the use of ionizing radiation, especially X-rays, in medical diagnosis.

rare earth element the metallic elements found in the two bottom rows of the periodic table; the chemistry of their ions is determined by electronic configurations with partially filled "f" subshells; see **lanthanides** and **actinides.**

reactant the chemical species present at the beginning of a chemical reaction that rearrange atoms to form new species.

reducing agent a chemical reagent that causes an element in another reagent to be reduced to a lower oxidation state.

reduction a decrease in an atom's oxidation state; accomplished by a gain of electrons or a decrease in the number of chemical bonds to atoms of other elements; see **oxidation state.**

r-process the rapid capture by iron nuclei of a succession of neutrons, occurring during supernova explosions.

sandstone a sedimentary geologic deposit composed mostly of sand-size grains, usually quartz, combined with other cementing material.

sea of instability on a chart of nuclides on which numbers of protons are graphed in relation to numbers of neutrons, the locations of atomic nuclei that have half-lives shorter than one second.

sedimentary a geologic term referring to rock formed from sediments accumulated over time.

semimetal see **metalloid.**

sequestration a means of isolating a material, often discussed with regard to carbon dioxide.

shell all of the electron orbitals that have the same value of the principal energy level, notated as "n".

shield for inner shell electrons, refers to the inability of external electric fields to penetrate the outer electron barrier.

Sodium-Cooled Fast Reactor a nuclear reactor that uses fast neutrons to produce the fissile fuel Pu-239 from U-238. Sodium, rather than water, is used as coolant because it does not significantly slow the neutrons.

solute a substance present in lesser amount in a solution. A solution can have one or several solutes; for example, in seawater, each dissolved salt or gas is a solute.

solubility the extent to which a solute may dissolve in a solvent at a particular temperature.

solvent the substance present in greatest amount in a solution; in *aqueous* solutions, water is the solvent.

specific heat the heat per unit mass needed to raise the temperature of a substance by one degree.

spectral lines narrow lines of light or dark in a spectrum, caused by discrete electron energy changes in a gas.

spectrum the range of electromagnetic radiation arranged in order of wavelengths or frequencies; for example, the visible spectrum, in order of increasing frequency exists in the order red, orange, yellow, green, blue, indigo, and violet.

spontaneous fission the fission of a nucleus without the event having been initiated by external causes.

s-process in massive stars, the so-called slow process, in which nuclei with masses greater than or equal to that of iron absorb relatively slow neutrons over long periods of time to form heavier elements.

stalactite a rock formation that precipitates from the ceiling of a limestone cave.

stalagmite a rock formation that precipitates on the floor of a limestone cave.

subatomic particle a particle that is smaller than an atom.

subshell all of the orbitals of a principal shell that lie at the same energy level.

superconductor a material in which current flow experiences virtually no resistance as it travels through the material, so no power is lost as heat.

supercriticality the condition in which a nuclear chain reaction proceeds at a rate sufficiently rapid to cause an explosion.

supernova a colossal explosive event ending the evolution of a high-mass star and ejecting its matter into interstellar space.

tailings debris or waste material from mining operations.

tetravalent describing an atom that can form four chemical bonds.

tracer in medicine, a radioactive substance used to image problems such as tumors; in oceanography used to image water mass movement.

transactinide an element that comes after lawrencium (element 103) in the periodic table.

transistor an electronic circuit device used to modify signals.

transition metal any of the metallic elements found in the 10 middle columns of the periodic table to the right of the alkaline earth metals; the chemistry of their ions largely is determined by electronic configurations with partially filled d subshells.

transmutation the conversion by way of a nuclear reaction of one element into another element; in transmutation, the atomic number of the element must change.

transcurium element any element in the periodic table with an atomic number greater than 96 (the atomic number of curium).

transfermium element any element in the periodic table with an atomic number greater than 100 (the atomic number of fermium).

transplutonium element any element in the periodic table with an atomic number greater than 94 (the atomic number of plutonium).

transuranium element any element in the periodic table with an atomic number greater than 92.

triad a group of three elements that have very similar chemical and physical properties; an example is sulfur, selenium, and tellurium.

trioxide an inorganic compound that contains three oxygen atoms.

unconformity in geology, referring to sedimentary depositions between hard rock layers with different ages.

uranium enrichment see **enrichment.**

vein in geology, referring to a long, narrow, rich deposit of a particular ore.

Very High Temperature Reactor a graphite-moderated, helium-cooled reactor design that promises increased efficiency.

X-ray very short wavelength, high-frequency electromagnetic radiation; falls between ultraviolet light and gamma rays in frequency.

yellowcake the end product of the uranium mining process, consisting of the usable uranium oxide that has been separated from other minerals; usually brownish in color.

yttria subgroup the lanthanides that tend to be found in ores with yttrium.

Further Resources

THE LANTHANIDES OR RARE EARTH ELEMENTS

Books and Articles

Dent, Peter C. "High Performance Magnet Materials: Risky Supply Chain." *Advanced Materials & Processes* (August 2009): 27–30. This article discusses the technological uses of rare earth materials and the need to find sources beyond China.

Evans, C. H., ed. *Episodes from the History of the Rare Earth Elements.* Dordrecht, The Netherlands: Kluwer Academic Publishers, 1996. An excellent short book about the people and events that led to the discoveries of the rare earth elements; in addition, some of the geology of the rare earth elements and their modern applications.

Gschneidner, Karl A., Jr., and Eyring, Le Roy, eds. *Handbook on the Physics and Chemistry of Rare Earths.* Amsterdam: North-Holland Physics Publishing, 1979. A comprehensive treatment of the properties of the rare earth elements.

Jensen, William B. "The Positions of Lanthanum (Actinium) and Lutetium (Lawrencium) in the Periodic Table." *Journal of Chemical Education* 59, no. 8 (1982): 634–636. A summary of the debate whether to place lanthanum, actinium, lutetium, and lawrencium in Group IIIA under scandium and yttrium, or to place them in the lanthanide and actinide rows.

Lavelle, Laurence. "Lanthanum (La) and Actinium (Ac) Should Remain in the d-Block." *Journal of Chemical Education* 85, no. 11 (2008): 1,482–1,483. An argument for positioning lanthanum and actinium under scandium and yttrium in the periodic table rather than in the lanthanide and actinide rows.

Moeller, Therald. "Periodicity and the Lanthanides and Actinides." *Journal of Chemical Education* 47, no. 6 (1970): 417–423. A explanation of the differences in the periodic properties of "f"-block elements as opposed to elements in the other blocks of the periodic table.

Pizzi, Richard A. "Jons Jakob Berzelius." *Today's Chemist at Work* 13, no. 12 (2004): 54–57. A brief summary of the life and achievements of the famous Swedish chemist Jöns J. Berzelius.

Service, Robert. "Is Silicon's Reign Near Its End?" *Science* 323 (February 2009): 1,001–1,002. This article gives information about the possible roles of lanthanum and lutetium in the semiconductor industry.

Ternström, Torolf. "Subclassification of Lanthanides and Actinides." *Journal of Chemical Education* 53, no. 10 (1976): 629–631. An argument to group the lanthanides and actinides in ways that would better demonstrate their similarities to other elements in the periodic table.

Internet

Bradsher, Keith. "Earth-Friendly Elements, Mined Destructively." Available online. URL: www.nytimes.com/2009/12/26/business/global/26rare.html. Accessed January 1, 2010. Discusses the environmental costs of rare earth mining, focusing on Chinese dysprosium and terbium mining.

Margonelli, Lisa. "Clean Energy's Dirty Little Secret." Available online. URL: www.theatlantic.com/doc/200905/hybrid-cars-minerals. Accessed December 15, 2009. This article discusses the environmental costs of rare earth mining, focusing on the Mountain Pass mine in California.

United States Geological Survey. "Rare Earths." Available online. URL: minerals.usgs.gov/minerals/pubs/commodity/rare_earths/mcs-2009-raree.pdf. Accessed December 18, 2009. The latest USGS information on rare earth mineral concentrates as a world commodity, including production and use, prices, events, trends, and issues.

———. "Rare Earths—Critical Resources for High Technology." Available online. URL: pubs.usgs.gov/fs/2002/fs087-02/. Accessed December 18, 2009. This Web article discusses the uses in current technology of the rare earth minerals.

Zimmerman, Martin. "California Metal Mine Regains Luster." Available online. URL: articles.latimes.com/2009/oct/14/business/fi-rare-

earth14. Accessed December 15, 2009. This article discusses the environmental costs of rare earth mining, focusing on the Mountain Pass mine in California.

ACTINIUM, THORIUM, AND PROTACTINIUM

Books and Articles

Dauphas, Nicolas. "The U/Th Production Ratio and the Age of the Milky Way from Meteorites and Galactic Halo Stars." *Nature* 435 (June 30, 2005): 1,203–1,205. This article discusses how knowing the uranium-to-thorium ratio of abundances in meteorites can help give an age for the galaxy.

Sime, Ruth Lewin. "The Discovery of Protactinium." *Journal of Chemical Education* 63, no. 8 (1986): 653–657. An account of the events that culminated in the discovery or protactinium.

Waggoner, William H. "The First Actinium Claim." *Journal of Chemical Education* 53, no. 9 (1976): 580. An account of a claim for the discovery of a new element almost 20 years prior to the actual study of actinium—the term *actinium* was proposed for the earlier element, but its existence was never confirmed.

Internet

Asaravala, Amit. "Thorium Fuels Safer Reactor Hopes." Available online. URL: www.wired.com/science/discoveries/news/2005/07/68045. Accessed January 4, 2010. This article discusses the environmental costs of rare earth mining, focusing on the Mountain Pass mine in California.

Bryan, A. Canon. "Thorium as a Secure Nuclear Fuel Alternative." Available online. URL: http://www.ensec.org/index.php?option=com_content&view=article &id=187:thorium-as-a-secure-nuclear-fuel-alternative&catid=94:0409content&Itemid=342. Accessed May 26, 2010. This article from the Institute for the Analysis of Global Security details the pros and cons of thorium fission reactors.

Environmental Protection Agency. "Thorium." Available online. URL: www.epa.gov/rpdweb00/radionuclides/thorium.html. Accessed January 25, 2010. This site describes thorium properties, uses, and environmental effects.

URANIUM

Books and Articles

Brugge, Doug, Timothy Benally, and Esther Yazzie-Lewis, eds. *The Navajo People and Uranium Mining.* Albuquerque: University of New Mexico Press, 2006. This accessible book tells about the experiences of Navajo miners and families in the uranium mining industry.

McDonald, Avril, Jann K. Kleffner, and Brigit C. A. Toebes, eds. *Depleted Uranium Weapons and International Law: A Precautionary Approach.* Cambridge: Cambridge University Press, 2008. This important book informs the reader about the history and hazards of depleted uranium and suggests solutions.

Meshik, Alex P. "The Workings of an Ancient Nuclear Reactor." *Scientific American* 293, no. 5 (November 2005): 82–91. This article gives details about an ancient uranium deposit that spontaneously became a nuclear reactor.

Settle, Frank A. "Uranium to Electricity: The Chemistry of the Nuclear Fuel Cycle." *Journal of Chemical Education* 86, no. 3 (2009): 316–323. A comprehensive but nontechnical explanation of the processes of mining and refining uranium ore, enriching the isotope mixture, fabricating the fuel elements, running the fission processing, and reprocessing the spent fuel elements.

Sime, Ruth Lewin. "Lise Meitner and the Discovery of Fission." *Journal of Chemical Education* 66, no. 5 (1989): 373–376. Written on the occasion of the 50th anniversary of the discovery of fission, the author explains the political climate under which Lise Meitner worked and the events that led to fission's discovery.

Younger, Stephen M. *The Bomb.* New York: HarperCollins, 2009. This book guides the reader from the Manhattan Project to the cold war and into present-day issues of nuclear proliferation.

Zoellner, Tom. *Uranium: War, Energy, and the Rock That Changed the World.* London: Viking Penguin, 2009. This is a highly readable account of how the use of uranium in World War II literally changed society.

Internet

Bielo, David. "Finding Fissile Fuel." Available online. URL: www.scientificamerican.com/article.cfm?id=finding-fissile-fuel. Accessed January 15, 2010. This article discusses currently known uranium resources as well as the need to locate new sources owing to increased demand.

Environmental Protection Agency. "Uranium." Available online. URL: www.epa.gov/rpdweb00/radionuclides/uranium.html. Accessed January 25, 2010. This site describes uranium properties, uses, and environmental effects.

Franzen, Harald. "The Science of the Silver Bullet." Available online. URL: www.scientificamerican.com/article.cfm?id=the-science-of-the-silver. Accessed January 26, 2010. This article discusses the use of depleted uranium in weapons and armor and the possible negative consequences of this practice.

U.S. Nuclear Regulatory Commission. "New Reactors." Available online. URL: www.nrc.gov/reactors/new-reactors.html. Accessed January 19, 2010. This Web site details what aspects and in what ways this body regulates nuclear reactor facilities.

World Nuclear Association. "Geology of Uranium Deposits." Available online. URL: www.world-nuclear.org/info/inf26.html. Accessed January 9, 2010. This January 2010 report details where and how uranium occurs in soil and mineral deposits.

———. "World Uranium Mining." Available online. URL: www.world-nuclear.org/info/inf23.html. Accessed January 16, 2010. A November 2009 report that details which countries have the highest uranium production.

THE TRANSURANIUM ELEMENTS

Books and Articles

Bernstein, Jeremy. *Plutonium: A History of the World's Most Dangerous Element.* Ithaca, N.Y.: Cornell University Press, 2007. This book describes the history of the discovery of plutonium and its subsequent effects in well-written detail.

Cunningham, B. B. "Berkelium and Californium." *Journal of Chemical Education* 36, no. 1 (1959): 32–37. An early summary of the chemistry and isotopes of berkelium and californium by one of the principal investigators in the field of transuranium elements.

Hindman, James C. "Neptunium and Plutonium." *Journal of Chemical Education* 36, no. 1 (1959): 22–26. A summary of the chemical properties of neptunium and plutonium.

Keenan, T. K. "Americium and Curium," *Journal of Chemical Education* 36, no. 1 (1945): 27–31. A summary of the chemistry of americium and curium.

Kostecka, Keith. "Americium—From Discovery to the Smoke Detector and Beyond." *Bulletin for the History of Chemistry* 33, no. 2 (2008): 89–93. A good summary of the discovery, properties, applications, and chemistry of americium.

Medvedev, Zhores A. *The Legacy of Chernobyl.* New York: W. W. Norton, 1990. A comprehensive analysis of Chernobyl's long-term global effects on the environment, human health, and the economy.

Navratil, James D., Wallace W. Schulz, and Glenn T. Seaborg. "The Most Useful Actinide Isotope: Americium-241." *Journal of Chemical Education* 67, no. 1 (1990): 15–16. A summary of americium's chemistry, methods for purifying it, and applications.

Seaborg, Glenn T. "The Transcalifornium Elements." *Journal of Chemical Education* 36, no. 1 (1959): 38–44. A review of the discoveries of einsteinium, fermium, mendelevium, and nobelium by one of the principal discoverers.

Welsome, Eileen. *The Plutonium Files: America's Secret Medical Experiments in the Cold War.* New York: Random House, 2000. This book details the radiation experiments on humans carried out by the government in the 1940s.

Internet

Argonne National Laboratory. "Americium." Available online. URL: www.ead.anl.gov/pub/doc/Americium.pdf. Accessed January 25, 2010. This human health fact sheet describes how americium is produced and its properties, uses, and environmental and health effects.

Barras, Colin. "Earliest weapons-grade plutonium found in US dump." Available online. URL: www.newscientist.com/article/dn16447-earliest-weaponsgrade-plutonium-found-in-us-dump.html. Accessed January 27, 2010. This article describes the recent discovery of Pu-239 in an unexpected location.

Environmental Protection Agency. "Americium." Available online. URL: www.epa.gov/rpdweb00/radionuclides/americium.html. Accessed January 24, 2010. This site describes americium properties, uses, and environmental effects.

Federation of American Scientists. "Plutonium Production." Available online. URL: www.fas.org/nuke/intro/nuke/plutonium.htm. Accessed January 27, 2010. This site describes how plutonium is produced for use in weapons.

Union of Concerned Scientists. "Nuclear Weapons: How They Work." Available online. URL: www.ucsusa.org/assets/documents/nwgs/nuclearweaponshowtheyworkfinal.pdf. Accessed January 27, 2010. This document is a brief, readable overview of the mechanics of nuclear weapons.

World Nuclear Association. "Smoke Detectors and Americium." Available online. URL: www.world-nuclear.org/info/inf57.html. Accessed January 22, 2010. This site gives the basics about the use of americium in smoke detectors.

THE TRANSACTINIDES

Books and Articles

Hofmann, S. "Welcome Copernicium?" *Nature Chemistry* 2, no. 2 (February, 2010): 146. A review by the discoverer of element 112 of the discovery and subsequent recommendation that element 112 be named *copernicium.*

———, and G. Münzenberg. "The Discovery of the Heaviest Elements." *Reviews of Modern Physics* 72, no. 3 (2000): 733–767. Written by the discoverers of several of the transactinide elements, this article is a comprehensive technical review of the discoveries.

Koppenol, W. H. "Naming of New Elements." *Pure and Applied Chemistry* 74, no. 5 (2002): 787–791. Contains the IUPAC recommendations for procedures for naming new elements.

Minkle, J. R. "Element 118 Discovered Again—for the First Time." *Scientific American* (October 17, 2006). This article describes how scientists in California and Russia synthesized element 118.

Peterson, I. "Element 112 Debuts in Fusion of Lead, Zinc." *Science News* 149, no. 9 (1996): 134. One of the first public announcements of the discovery of element 112.

Internet

Glanz, James. "Scientists Discover Heavy New Element." Available online. URL: http://www.nytimes.com/2010/04/07/science/07element.html. Accessed May 20, 2010. This article discusses the discovery by Russian and U.S. scientists of element 117.

Lemonick, Michael. "The Birth of a New Element." Available online. URL: http://www.time.com/time/health/article/0,8599,1546747,00.html. Accessed on February 10, 2010. A report from 2006 of the discovery of element 118.

Moody, Ken. Lawrence Livermore National Laboratory. "Present at the Creation." Available online. URL: https://www.llnl.gov/str/JanFeb02/Moody.html. Accessed on June 30, 2009. A discussion of the creation of elements 114 and 116 and what their discoveries suggest in regards to the search for superheavy elements.

A Time. "Nuclear Physics: The Heaviest Atom." Available online: URL: http://search.time.com/results.html?N=0&Nty=1&p=0&cmd=tags&srchCat=Full+Archive&Ntt=the+heaviest+atom. Accessed on February 10, 2010. A report from 1967 of the discovery of what was then the heaviest known atom—mendelevium 258.

FUTURE DIRECTIONS

Books and Articles

Clifton, Timothy, and Pedro G. Ferreira. "Does Dark Energy Really Exist?" *Scientific American* 300, no. 4 (2009): 48–55. An up-to-date review of the thinking about the nature of dark energy.

Karel, Paul J. "The Mendeleev–Seaborg Periodic Table: Through Z = 1138 and Beyond." *Journal of Chemical Education* 79, no. 1 (2002): 60–63. A suggestion of how to extend the periodic table if "g," "h," "i," etc., blocks are ever needed.

Seaborg, Glenn T. "Prospects for Further Considerable Extension of the Periodic Table." *Journal of Chemical Education* 46, no. 10 (1969): 626–634. In this article, Dr. Seaborg makes predictions of the chemical properties and electronic configurations of transactinide elements.

Internet

PBS. "Island of Stability." Available online: URL: www.pbs.org/wgbh/nova/sciencenow/3313/02.html. Accessed on February 10, 2010. A PBS video about the search for element 114.

Sciencedaily "Periodic Table: Nuclear Scientists Eye Future Landfall On A Second 'Island Of Stability.'" Available online: URL: www.sciencedaily.com/releases/2008/04/080406114739.htm. A report of attempts to synthesize long-lived superheavy elements.

General Resources

The following sources discuss general information on the periodic table of the elements.

Books and Articles

Ball, Philip. *The Elements: A Very Short Introduction.* New York: Oxford University Press, 2004. This book contains useful information about the elements in general.

Chemical and Engineering News 87, no. 29 (July 6, 2009): 51–59. A production index published annually showing the quantities of various chemicals that are manufactured in the United States and other countries.

Considine, Glenn D., ed. *Van Nostrand's Encyclopedia of Chemistry,* 5th ed. New York: John Wiley, 2005. In addition to its coverage of traditional topics in chemistry, the encyclopedia has articles on nanotechnology, fuel cell technology, green chemistry, forensic chemistry, materials chemistry, and other areas of chemistry important to science and technology.

Cotton, F. Albert, Geoffrey Wilkinson, and Paul L. Gaus. *Basic Inorganic Chemistry,* 3d ed. New York: John Wiley, 1995. Written for a beginning course in inorganic chemistry, this book presents information about individual elements.

Cox, P. A. *The Elements on Earth: Inorganic Chemistry in the Environment.* New York: Oxford University Press, 1995. There are two parts to this book. The first part describes Earth and its geology and how elements and compounds are found in the environment. Also, it describes how elements are extracted from the environment. The second part describes the sources and properties of the individual elements.

Daintith, John, ed. *The Facts On File Dictionary of Chemistry,* 4th ed. New York: Facts On File, 2005. Definitions of many of the technical terms used by chemists.

Downs, A. J., ed. *Chemistry of Aluminium, Gallium, Indium and Thallium.* New York: Springer, 1993. A detailed, wide-ranging, authoritative and up-to-date review of the chemistry of aluminium, gallium, indium, and thallium. Coverage is of the chemistry and commercial aspects of the elements themselves; emphasis is on the design and synthesis of materials, their properties and applications.

Emsley, John. *Nature's Building Blocks: An A-Z Guide to the Elements.* New York: Oxford University Press, 2001. Proceeding through the periodic table in alphabetical order of the elements, Emsley describes each element's important properties, biological and medical roles, and importance in history and the economy.

———. *The Elements.* New York: Oxford University Press, 1989. Emsley provides a quick reference guide to the chemical, physical, nuclear, and electron shell properties of each of the elements.

Foundations of Chemistry 12, no. 1 (April 10, 2010). This special issue of the journal focuses on the periodic table, featuring some obscure history, possible new arrangement of elements, and the role of chemical triads.

Gray, Theodore. *The Elements: A Visual Exploration of Every Known Atom in the Universe.* New York: Black Dog and Leventhal Publishers, 2009. A book packed with pictures of all the elements, including summaries of their properties and applications.

Greenberg, Arthur. *Chemistry: Decade by Decade.* New York: Facts On File, 2007. An excellent book that highlights by decade the important events that occurred in chemistry during the 20th century.

Greenwood, N. N., and A. Earnshaw. *Chemistry of the Elements.* Oxford: Pergamon Press, 1984. This book is a comprehensive treatment of the chemistry of the elements.

Hall, Nina, ed. *The New Chemistry.* Cambridge: Cambridge University Press, 2000. Contains chapters devoted to the properties of metals and electrochemical energy conversion.

Hampel, Clifford A., ed. *The Encyclopedia of the Chemical Elements.* New York: Reinhold, 1968. In addition to articles about individual elements, this book also features articles about general topics in chemistry. Numerous authors contributed to this book, all of whom were experts in their respective fields.

Heiserman, David L. *Exploring Chemical Elements and Their Compounds.* Blue Ridge Summit, Pa.: Tab Books, 1992. This book is described by its author as "a guided tour of the periodic table for ages 12 and up," and is written at a level that is very readable for precollege students.

Henderson, William. *Main Group Chemistry.* Cambridge: The Royal Society of Chemistry, 2002. This book is a summary of inorganic chemistry in which the elements are grouped by families.

Jolly, William L. *The Chemistry of the Non-Metals.* Englewood Cliffs, N.J.: Prentice-Hall, 1966. This book is an introduction to the chemistry of the nonmetals, including the elements covered in this book.

King, R. Bruce. *Inorganic Chemistry of Main Group Elements.* New York: Wiley-VCH, 1995. This book describes the chemistry of the elements in the "s" and "p" blocks.

Krebs, Robert E. *The History and Use of Our Earth's Chemical Elements: A Reference Guide,* 2d ed. Westport, Conn.: Greenwood Press, 2006. Following brief introductions to the history of chemistry and atomic structure, Krebs proceeds to discuss the chemical and physical properties of the elements group (column) by group. In addition, he describes the history of each element and current uses.

Lide, David R., ed. *CRC Handbook of Chemistry and Physics,* 89th ed. Boca Raton, Fla.: CRC Press, 2008. The *CRC Handbook* has been the most authoritative, up-to-date source of scientific data for almost nine decades.

Mendeleev, Dmitri Ivanovich. *Mendeleev on the Periodic Law: Selected Writings, 1869–1905.* Mineola, N.Y.: Dover, 2005. This English translation of 13 of Mendeleev's historic articles is the first easily accessible source of his major writings.

Norman, Nicolas C. *Periodicity and the p-Block Elements.* New York: Oxford University Press, 1994. This book describes group properties of post-transition metals, metalloids, and nonmetals.

Parker, Sybil P., ed. *McGraw-Hill Encyclopedia of Chemistry,* 2d ed. New York: McGraw Hill, 1993. This book presents a comprehensive treatment of the chemical elements and related topics in chemistry, including expert-authored coverage of analytical chemistry, biochemistry, inorganic chemistry, physical chemistry, and polymer chemistry.

Rouvray, Dennis H., and R. Bruce King, eds. *The Periodic Table: Into the 21st Century.* Baldock, Hertfordshire, U.K.: Research Studies Press Ltd., 2004. A presentation of what is happening currently in the world of chemistry.

Stwertka, Albert. *A Guide to the Elements,* 2d ed. New York: Oxford University Press, 2002. This book explains some of the basic concepts of chemistry and traces the history and development of the periodic table of the elements in clear, nontechnical language.

Winter, Mark J., and John E. Andrew. *Foundations of Inorganic Chemistry.* New York: Oxford University Press, 2000. This book presents an elementary introduction to atomic structure, the periodic table, chemical bonding, oxidation and reduction, and the chemistry of the elements in the "s," "p," and "d" blocks; in addition, there is a separate chapter devoted just to the chemical and physical properties of hydrogen.

Internet Resources

About.com: Chemistry. Available online. URL: chemistry.about.com/od/chemistryfaqs/f/element.htm. Accessed December 4, 2009. Information about the periodic table, the elements, and chemistry in general from the New York Times Company.

American Chemical Society. Available online. URL: portal.acs.org/portal/acs/corg/content. Accessed December 4, 2009. Many educational resources are available here.

Center for Science and Engineering Education, Lawrence Berkeley Laboratory, Berkeley, California. Available online. URL: www.lbl.gov/Education. Accessed December 4, 2009. Contains educational resources in biology, chemistry, physics, and astronomy.

Chemical Education Digital Library. Available online. URL: www.chemeddl.org/index.html. Accessed Accessed December 4, 2009. Digital content intended for chemical science education.

Chemical Elements. Available online. URL: www.chemistryexplained.com/elements. Accessed December 4, 2009. Information about each of the chemical elements.

Chemical Elements.com. Available online. URL: www.chemicalelements.com. Accessed December 4, 2009. A private site that originated with a school science fair project.

Chemicool. Available online. URL: www.chemicool.com. Accessed December 4, 2009. Information about the periodic table and the chemical elements, created by David D. Hsu. of the Massachusetts Institute of Technology.

Department of Chemistry, University of Nottingham, United Kingdom. Available online. URL: www.periodicvideos.com. Accessed December 4, 2009. Short videos on all of the elements can be viewed.

Journal of Chemical Education, Division of Chemical Education, American Chemical Society. Available online. URL: jchemed.chem.wisc.edu/HS/index.html. Accessed December 4, 2009. The Web site for the premier online journal in chemical education.

Lenntech. Available online. URL: www.lenntech.com/Periodic-chart.htm. Accessed December 4, 2009. Contains an interactive, printable version of the periodic table.

Los Alamos National Laboratory. Available online. URL: periodic.lanl.gov/default.htm. Accessed December 4, 2009. A resource on the periodic table for elementary, middle school, and high school students.

Mineral Information Institute. Available online. URL: www.mii.org. Accessed December 4, 2009. A large amount of information for teachers and students about rocks and minerals and the mining industry.

National Nuclear Data Center, Brookhaven National Laboratory. Available online. URL: www.nndc.bnl.gov/content/HistoryOf

Elements.html. Accessed December 4, 2009. A worldwide resource for nuclear data.

Periodic Table of Comic Books, Department of Chemistry, University of Kentucky. Available online. URL: www.uky.edu/Projects/Chemcomics. Accessed December 8, 2009. A fun, interactive version of the periodic table.

Royal Society of Chemistry. http://www.rsc.org/chemsoc/. Accessed December 4, 2009. This site contains information about many aspects of the periodic table of the elements.

Schmidel & Wojcik: Web Weavers. Available online. URL: quizhub.com/quiz/f-elements.cfm. Accessed December 4, 2009. A K-12 interactive learning center that features educational quiz games for English language arts, mathematics, geography, history, earth science, biology, chemistry, and physics.

United States Geological Survey. Available online. URL: minerals.usgs.gov. Accessed December 4, 2009. The official Web site of the Mineral Resources Program.

Web Elements, The University of Sheffield, United Kingdom. Available online. URL: www.webelements.com/index.html. Accessed December 4, 2009. A vast amount of information about the chemical elements.

Wolfram Science. Available online. URL: demonstrations.wolfram.com/PropertiesOfChemicalElements. Accessed December 4, 2009. Information about the chemical elements from the Wolfram Demonstration Project.

Index

Note: *Italic* page numbers refer to illustrations.

A

B

C

H

I

J